TRADITIONAL WEAPONS OF THE INDONESIAN ARCHIPELAGO

A. **Ceremonial knife**
Bali, Bulelang. RMV 1050-2. Acq. in 1895 from mr. L.A.J.W. Sloet Van Der Beele Van Nispen Heirs.

B. **Katapu**
Kalimantan. RMV 401-40. Acq. in 1883 from mr. J.W. Van Lansberge. H. 16.5 cm.

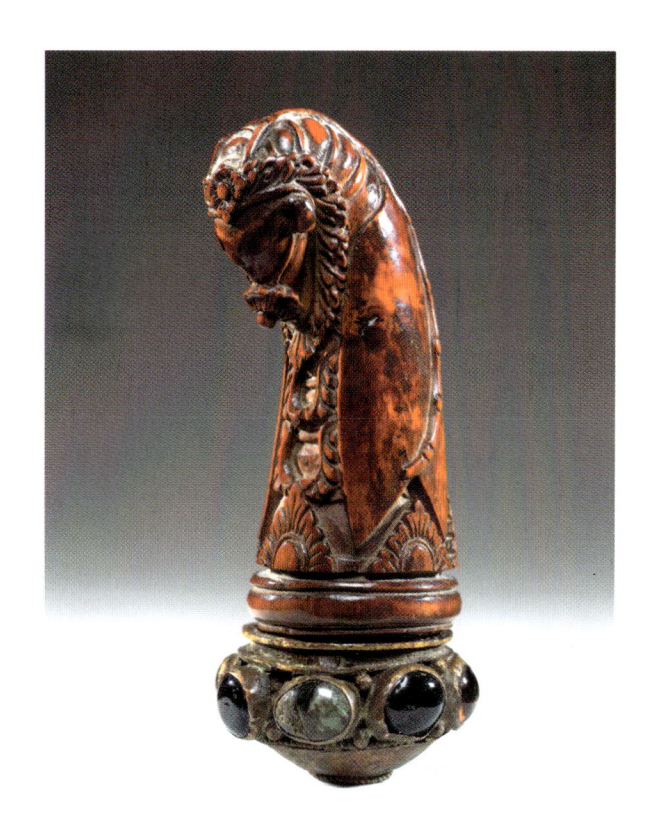

C. **Keris hilt, Kocet-Kocetan**
Bali. Wood, copper, glass. H. 10.2 cm.

D. **Keris panjang**
Sumatra. Hilt of a fossil molar of an elephant.

This book is dedicated to Joke, Kanta and Ivo.

E. **Tolor**
Kalimantan. RMV 2292-93. Acq. in 1935 from G.L. Krol.
H. 31.5 cm.

F. **Pedang lurus**
Java. RMV 1848-3. Acq. in 1913 from dr. A.W. Pulle.

TRADITIONAL WEAPONS
OF THE INDONESIAN ARCHIPELAGO

by ALBERT G. VAN ZONNEVELD

C. ZWARTENKOT ART BOOKS - LEIDEN

ACKNOWLEDGEMENTS

This book is the result of enthusiastic contributors. Their knowledge and skill, their views and advice, and especially their continuous support, has resulted in this reference work dealing with a very important segment of the material culture of the Indonesian archipelago.
Firstly I must express my gratitude to the publisher for his ongoing assistance throughout the project. He was also responsible for digitally reproducing a large number of objects and plates included in this publication. Furthermore, my thanks are due for the valuable advice received from: Nico de Jonge, visiting keeper at the Rijksmuseum voor Volkenkunde (Leiden) and a specialist in the art and culture of Maluku as well as other East Indonesian regions; David van Duuren, author on the keris and attached to the Tropenmuseum (Amsterdam); David Stuart-Fox, librarian of the above-mentioned Leiden museum, who shared his knowledge of local Indonesian literature. The translation was presented by Peter Richardus in collaboration with Timothy D. Rogers of the Bodleian Library (Oxford) and author on Indonesian weapons also.
Bruce van Rijk of the Instituut Indonesische Cursussen (Leiden) gave advice on the contemporary spelling of Indonesian terms.
I am also indebted to the Management Team of the Rijksmuseum voor Volkenkunde for allowing me to publish a number of masterpieces.
Ben Grishaaver and Eleonora van der Bij saw to it that they were excellently photographed. They were assisted by Dick Speksnijder and Andy Smith, both of the M-Company. Ernest Boesaard gave skillful advice regarding digital photography. The plates depicting the keris hilts from the Crince Le Roy Collection were made by Ferry Herrebrugh. Concept, design and production are by Allard de Rooi and Maarten van der Kroft who have created several fine books on Indonesia.
Needless to say, the participation of collectors was of utmost importance to the realization of this book. These contributors follow, in alphabetical order. Marion and Eric Crince Le Roy kindly presented a selection from their many, excellent keris hilts; Willem van der Post, having extensively worked on this project, drew from his large collection of weapons and shields; Karel Sirag's excellent collection of weapons is partly presented as well as his description of Timorese and Tanimbarese swords and Nias sword hilts. In the course of time a large number of cognoscenti and collectors active in this field kindly forwarded literature and information used in this publication. A special mention must here be made of Arnold Wentholt for allowing an unpublished manuscript to be used. It proved an important contribution to this book. I am also indebted to Mr. J. van Daalen en Mrs. Iris M. van Daalen for allowing me to publish a selection of their kerisses.
Last but not least my gratitude goes out to Emmar van Duin for her generosity and mental support throughout the project.

Albert van Zonneveld
Leiden, August 2001

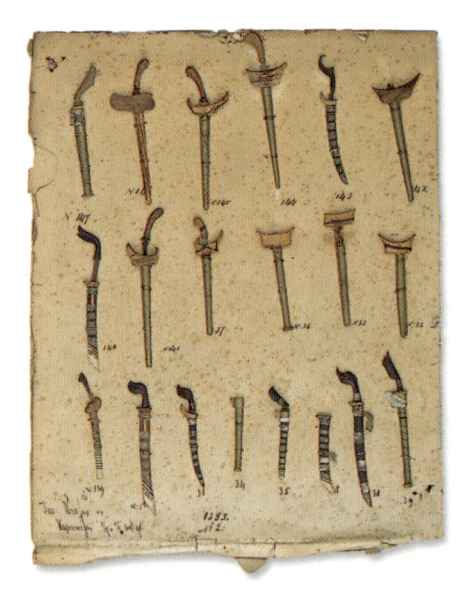

G. **Tableau miniature weapons**
RMV 1595-2. Acq. in 1907 from dr. J.W.L. Goethart.

COLOPHON

Published by C. Zwartenkot, Art Books - Leiden
© 2001 A.G. van Zonneveld and the publisher

ALL RIGHTS RESERVED

No part of this book may be reproduced in any form without written permission from the publisher

Printed in The Netherlands

ISBN 90-5450-004-2
NUGI 648

Publisher: C. Zwartenkot
Translation: P. Richardus and T.D. Rogers
Design and production:
A. de Rooi and M. van der Kroft
The Ad Agency, Alphen aan den Rijn
Photography: B. Grishaaver, F. Herrebrugh
and the publisher
Jacket photography: B. Grishaaver
(exept Kayan warrior and amulet holder)
Colour separation: Telstar Prepress, Pijnacker
Printing: Scholma Druk, Bedum

THIS BOOK CAN BE OBTAINED FROM:
Ethnographic Art Books / Boekhandel De Verre Volken
c/o National Museum of Ethnology
Steenstraat 1
2312 BS Leiden
The Netherlands
Tel. + 31 (0)71 5168 706
Fax + 31 (0)71 5289 128
e-mail: info@ethnographicartbooks.com
website and online bookstore:
www.ethnographicartbooks.com

CONTENTS

Introduction	11
Map of Indonesia	12-15
Scheme	17-19
Alphabetical survey	21
A	22
B	26
CD	38
EFG	44
HIJ	50
K	58
L	78
MNO	86
P	94
R	112
S	116
T	140
UW	150
Index of names	154
Select bibliography	156
Provenance	160

H. Keris
Bali. RMV 3600-193. Acq. in 1959 through the Ethnografisch Museum van de Koninklijke Militaire Academie, Breda. L. 53 cm.

I. Kanta
Sulawesi. RMV 3600-5831. Acq. in 1959 through the Ethnografisch Museum van de Koninklijke Militaire Academie, Breda. L. 111.5 cm, W. 16 cm.

INTRODUCTION

Almost no other area in the world has produced such a varied arsenal of traditional weapons as the Indonesian archipelago. Laden with symbolism and blessed with divine power, the keris especially has achieved great fame. However, this vast region has brought forth numerous other functional and ritual weapons. Through the centuries an almost endless quantity of weapons and other implements of war have been produced by the people and groups living all over the archipelago.

Never before has this varied array of traditional weapons been so extensively listed and described. In 160 pages text and over 650 illustrations, numerous weapons - arranged alphabetically by name, with much information about shapes of blades and provenance - are opened up to professional purposes and to the devotee of the subject. Completeness was aimed at. However, both the endless variations of weapons and the fragmentary sources confined the research.

This reference work is limited to weapons essentially meant for warfare, including weapons with two uses: as a combat weapon and as an (agricultural) tool. Weapons exclusively designed for hunting and fishing are excluded. Decorations or insignia of war were left out also.

The geographical region covered by this book tallies mainly with present-day Indonesia. The weapons of Irian Jaya - politically a part of Indonesia, but ethnographically linked to Melanesia - have not been included. Weapons from Sarawak - politically a part of Malaysia, but ethnographically part of Kalimantan - were, however, included.

Regarding the spelling of names and terms, the present Indonesian spelling is followed as much as possible, with the exception of Java (Jawa) and Sumatra (Sumatera). For the names of other islands present-day Indonesian names are used, such as Kalimantan (Borneo), Sulawesi (Celebes) and Maluku (Moluccas).

Tarutao I.
Ladang I.
Langkawi
Yala
Alor Setar
Kota Baharu
Pulau We
Pulau Breueh
Banda Aceh
Pidië
Samalanga
Tanjung Jambuair
George Town
Pinang
Kuala Terengganu
Tasik Temengor
Kelantan
Perak
Muda
Tasik Kenyir
Lhokseumawe
Gayo
2590
Danau Lauttawar
Woyla
Langsa
Ujung Tamiang
Ipoh
ACEH
MALAYSIA
Tripa
Alas
3012
3145
Gunung Leuser
Binjai
Medan
Karo Batak
Tebingtinggi
Strait of Malacca
Kuantan
Pahang
Kelang
KUALA LUMPUR
Tanjung Dewa
Sembilan
Pematangsiantar
Tanjungbalai
Seremban
Tioman
Jemaja
Sidikalang
Asahan
Balige
Danau Toba
Sinabang
Simeulue
Labuhanbilik
SUMATERA
2300
Bagansiapiapi
Muar
Malacca
Tarutung
Rantauprapat
UTARA
Rupat
Johor Baharu
Tanjung Dowi
Sibolga
Rokan
Dumai
Bengkalis
SINGAPORE
SINGAPORE
Padangsidempuan
Gunungsitoli
Padang
Strait of Singapore
Rangsang
Tanjungpinang
RIAU
Rantau
Nias
Bawelowalani
Telukdalam
Pekanbaru
Kepulauan Riau
2172
Minangkabau
Kampar
2274
Pini
EQUATOR
2912
Gunung Ophir
Payakumbuh
Kepulauan Lingga
Tanahmasa
Bukittinggi
Danau Maninjau
Danau Singkarak
Rengar
Pulau-Pulau Batu
Padangpanjang
Singkep
Sawahlunto
Tanjung Sigep
Solok
Selat Berhala
Padang
SUMATERA
BARAT
Batanghari
Tanjung Jabung
2690
Simpang
Siberut
Selat Mentawai
Jambi
JAMBI
3805
Belinyu
Tembesi
Bangka
Sungaipenuh
Kubu
Pangkalpinang
Kepulauan Mentawai
Sipura
Selat Bangka
Koba
SUMATERA
Musi
Toboali
Pagai Utara
Palembang
Pagai Selatan
SELATAN
Pegunungan Barisan
2467
Bukit Duen
Mesuji
Tanjung Kait
Perabumulih
Bengkulu
Lahat
Tanjung Lumut
3159
Gunung Dempo
BENGKULU
Kotabumi
Sumatera
Sumatra
Danau Ranau
Tulangbawang
2232
Kotabumi
LAMPUNG
Enggano
Tanjungkarang-Telubketung
Tanjung Tua
Tanjung Cina
812
Krakatau
Selat Sunda
Serang
Jatine
Rangkasbitung
Bogor
Tanjung Guhakolak
Cianjur
Sukabumi
INDIAN OCEAN
0 km
80
160
240 km
1 JAKARTA
2 JAWA TENGAH
3 YOGYAKARTA
4 JAWA TIMUR
5 SULAWESI TENGAH

SOUTH CHINA SEA
4101
Gunung Kinabalu
Kota Kinabalu
Sandakan
Labuan
Brunei Bay
Kinabatangan
Tawitawi
BRUNEI
BANDAR SERI BEGAWAN
SABAH
Tanjong Baram
Miri
Tawau
Sebatik
Natuna Besar
Kepualan Natuna
Baram
Tarakan
Tarakan
CELEBES SEA
MALAYSIA
Subi Besar
Tanjong Sirik
Kayan
Tanjungselor
Rajang
2499
Selat Serasan
Sibu
SARAWAK
Tanjong Datu
Rajang
Tanjungredeb
Tanjung Batu
Maratua
Tanjung Munggguresak
2467
Liku
Baleh
2988
Pegunungan Iran
Kuching
KALIMANTAN TIMUR
1767 Gunung Lawit
Sambaliung
Singkawang
1701 Gunung Niyut
1960
2240 Bukit Liangpran
2000 Gunung Menyapa
Sangkulirang
510
Tanjung Mangkalihat
Tanjung Dondo
Balaikarangan
Landak
Mahakam
1730
Telen
1390
1744
KALIMANTAN BARAT
Pegunungan Muller
Pontianak
EQUATOR
Tanjung Manimbaya
Kapuas
Kenohan Jempang
Teluk Tomini
Telukpakedai
Borneo
Kalimantan
1758 Gunung Saran
Pulau Terentang
Samarinda
Donggala
Parigi
Kotabaru
Telugbatang
KALIMANTAN TENGAH
Muarateweh
Pulau Pemarung
Palu
Maya
Sandai
Kapuas
Balikpapan
Makassar Strait
3127
Palu
Kulawi
Selat Karimata
Karimata
Tanjung Bawang
Pawan
Kahayan
Mendawai
Barito
Kano
Bada
Sampit
Pembuang
Belitung
Belangiran
Palangkaraya
KALIMANTAN SELATAN
Karama
2950
Kendawangan
Amuntai
SULAWESI SELATAN
Tanjungpandan
Jelai
Sampit
Kandangan
Tanjung Rangas
Manggar
Tanjung Sambar
Kualakapuas
Pegunungan Meratus
1892 Gunung Besar
3074
Palopo
Makale
Tanjung Kluang
Banjarmasin
Martapura
Sebuku
Majene
3455 Gunung Rantemario
Tanjung Rangasa
Teluk Mandar
Pinrang
Laut
Parepare
Tanjung Puting
Batakan
Tanjung Layar
Danau Tempe
Sulawesi Celebes
JAVA SEA
1395
Watampone
Kepulauan Laut Kecil
Ujung Pandang
2871
Masalembu Besar
Bawean
Pulau-pulau Karimunjawa
Tanjung Indramayu
Indramayu
Tanjung Bugel
Salayar
Pemalang
Pekalongan
Pati
Rembang
Tanjung Pangkah
Madura
Kepulauan Sabalana
Cirebon
Kudus
Tuban
Arjasa
Kepulauan Kangean
Sumedang
3078
Brebes
Tegal
Batang
Demak
Blora
Bangkalan
Sumenep
Kuningan
Semarang
Bojonegoro
Pamekasan
Garut
3428 Gunung Slamet
Salatiga
Gresik
Sampang
Sepanjang
Ciamis
3135
Magelang
Ngawi
Mojokerto
Purwokerto
Madiun
Surabaya
Selat Madura
Cilacap
Surakarta
Sidoarjo
Kepulauan Tengah
Flores Sea
Purworejo
Pasuruan
Tanjung Pacinan
Kambagan
Ponorogo
Kediri
3343
Probolinggo
Situbondo
Bali
Bali Sea
Yogyakarta
Tulungagung
Malang
3088
Banyuwangi
Singaraja
Blitar
3676
Jember
Lumajang
3332
Selat Bali
BALI
3142
Selat Lombok
3726
Moyo
2821 Gunung Tambora
Sangeang
Jawa Java
Mataram
Teluk Saleh
Bima
Praya
Taliwang
Ruteng
Selat Sape
Semenanjung Blambangan
Denpasar
Lombok
Selat Alas
Sumbawa
Flores
NUSA TENGGARA BARAT
Selat Sumba
Karuni

CELEBES SEA
BANDA SEA
M A L
Molucca Sea
Maluku Moluccas
Makassar Strait
Flores Sea
Savu Sea
Timor Sea
Halmahera Sea
Seram
Sulawesi Celebes
SULAWESI UTARA
SULAWESI TENGAH
SULAWESI SELATAN
SULAWESI TENGGARA
KALIMANTAN TIMUR
NUSA TENGGARA BARAT
NUSA TENGGARA TIMUR
EAST TIMOR
Territory under transitional administration
EQUATOR
Tawau
Sebatik
Tarakan
Tanjungselor
Samarinda
Balikpapan
Sangkulirang
Manado
Tondano
Bolaang Mongondow
Kotamobagu
Gorontalo
Bilungala
Taludaa
Tolitoli
Palu
Donggala
Parigi
Poso
Luwuk
Kulawi
Tenteno
Matana
Usu
Wotu
Palopo
Makale
Majene
Pinrang
Parepare
Watampone
Ujung Pandang
Malamala
Kendari
Kolaka
Tampo
Ereke
Baubau
Teluk Tomini
Teluk Bone
Buton
Wowoni
Muna
Kabaena
Kepulauan Banggai
Peleng
Kepulauan Sula
Taliabu
Mangole
Sanana
Auponhia
Buru
Namlea
Ambon
Amahai
Obi
Halmahera
Morotai
Daruba
Galela
Tobelo
Kao
Kusu
Ternate
Tidore
Weda
Dolit
Labuna
Sangihe
Siau
Biaro
Karakelong
Kepulauan Tukangbesi
Salayar
Kepulauan Sabalana
Kepulauan Bonerate
Kalaotoa
Wetar
Selat Wetar
Kisar
Kepulauan Leti
Damar
Teun
Roma
Sumbawa
Moyo
Taliwang
Bima
Sangeang
Flores
Reo
Ruteng
Manggarai
Ende
Kewapante
Larantuka
Adonara
Lembata
Solor
Kepulauan Solor
Kepulauan Alor
Alor
Kalabahi
Timor
DILI
Baukau
Kupang
Roti
Baa
Savu
Sumba
Waingapu
Karuni
Selat Sumba

0 km
80
160
240 km
PACIFIC OCEAN
Kepulauan Mapia
Waigeo
EQUATOR
Tanjung Yamursba
Selat Dampir
3000 Kwoka
Karabra
Sorong
Mubrani
Tanjung Saweba
Andai
Numfor
Biak
2821
Jazirah Doberai
Teminabuan
Ransiki
Selat Yapen
Yapen
Tanjung Yamtu
Misoöl
Tanjung Saibi
1496
Serui
Tanjung d'Urville
Sarmi
Barma
Mogoi
Kamundan
Bintuni
Teluk Bintuni
Teluk Cenderawasih
New Guinea
Teluk Berau
Wasior
Mamberamo
Pegunungan Van Rees
Jayapura
Vanimo
Fakfak
Semanjung Bomberai
Napanwainami
Tariku
Bula
IRIAN JAYA
Hamuku
Nabire
Taritatu
3019
Kaimana
Derewo
Wandai
3892
Teluk Kamrau
Pegunungan Maoke
4234
Gunung Ubia
5030
Puncak Jaya
Wamena
4419
4730
Puncak Tricora
Sepik
Tanjung Papisoi
Tanjung Namaripi
Cemara
4595
Puncak Yamin
PAPUA NEW GUINEA
4170
Kepulauan Kai
Tanjung Borang
Pulau
Digul
Kai Besar
Tual
Wokam
Tanjung Weduar
Kobroör
Kepulauan Aru
Stickland
Trangan
Doka
Mapi
Tanjung Ngabordamlu
Tanjung Deyong
Digul
Murray Lake
Tanjung Dolak
Aramia
Watmuri
Fly
Yamdena
Pulau Dolak
Balimo
Samlaki
Kepulauan Tanimbar
Selaru
Tanjung Vals
Merauke
Morehead
Boigu I.
Saibai I.
Torres Strait
Badu I.
Moa I.
Prince of Wales I.
Cape York
Bamaga
Cape York
ARAFURA SEA

J. Keris
Kalimantan. RMV 970-23. Acq. in 1893. L. 66.5 cm.

CLASSIFICATION ACCORDING TO THE SHAPE OF THE BLADE

The blade's shape offers the best indication when determining which type of knife or sword we are dealing with. A scheme for the classification of edged weapons according to the blade's shape now follows. By determining the shape of the edge, the back, and the tip (and, for some types, other characteristics of the blade) step by step, the name can be ascertained for a number of weapons. Some blade forms are used for a number of different types of weapons. In that case the scheme refers to all these types. By looking up the various names in the Alphabetical Survey and comparing the information or plates with the weapon at hand, one can identify it in most cases.

THE FOLLOWING SHAPES CAN BE DISTINGUISHED:

EDGE
- **TYPE 1:** Straight or almost straight
- **TYPE 2:** Bent (convex)
- **TYPE 3:** Slightly curved (convex)
- **TYPE 4:** Clearly curved (convex)
- **TYPE 5:** Somewhat S-shaped
- **TYPE 6:** Slightly curved (concave)
- **TYPE 7:** Clearly curved (concave)
- **TYPE 8:** Straight or almost straight with a slightly concave part near the tip
- **TYPE 9:** Undulating

BACK
- **TYPE 1:** Straight or almost straight
- **TYPE 2:** Bent (concave)
- **TYPE 3:** Slightly curved (concave)
- **TYPE 4:** Clearly curved (concave)
- **TYPE 5:** Somewhat S-shaped
- **TYPE 6:** Slightly curved (convex)
- **TYPE 7:** Clearly curved (convex)
- **TYPE 8:** Non-applicable
- **TYPE 9:** Undulating

TIP
- **TYPE 1:** Slightly curved towards the back, ending in a point
- **TYPE 2:** Back curving towards the edge
- **TYPE 3:** Edge curving towards the back
- **TYPE 4:** Back running at an angle towards the edge
- **TYPE 5:** Edge running at an angle towards the back
- **TYPE 6:** Back and edge curving towards each other to a point
- **TYPE 7:** Back and edge curving towards each other to a sharp point
- **TYPE 8:** The tip of the blade runs straight or at an angle, and has one or more V-shaped indentations
- **TYPE 9:** The back near the tip is S-shaped, bent or concavely curved
- **TYPE 10:** The end is rectangular
- **TYPE 11:** The back runs at an angle towards the edge and has a spike or protrusion on the angled part

OTHER CHARACTERISTICS
The group of blades with a straight edge and back can be divided into three variants:
- **TYPE 1:** Edge and back run parallel
- **TYPE 2:** Blade broadens towards the tip
- **TYPE 3:** Blade narrows towards the tip

The group of blades of which the edge is clearly concave and the back clearly convex can also be divided into three variants:
- **TYPE 1:** Light, dagger-shaped blade
- **TYPE 2:** Light, crescent-shaped blade
- **TYPE 3:** Heavy blade

USE OF THE SCHEME
In the scheme the aforementioned shapes of edge, back and tip of the blade are indicated. When using the scheme, the weapon must be laid down with the hilt at the top, and the tip at the bottom. The back faces towards the left, the edge towards the right.
In the classification scheme are presented from top to bottom:

- the shape of the edge;
- the shape of the back;
- the remaining characteristics (if applicable);
- the shape of the tip;
- a reference to the names of the relevant weapons follows.

CLASSIFICATION SCHEME ACCORDING TO THE SHAPE OF THE BLADE

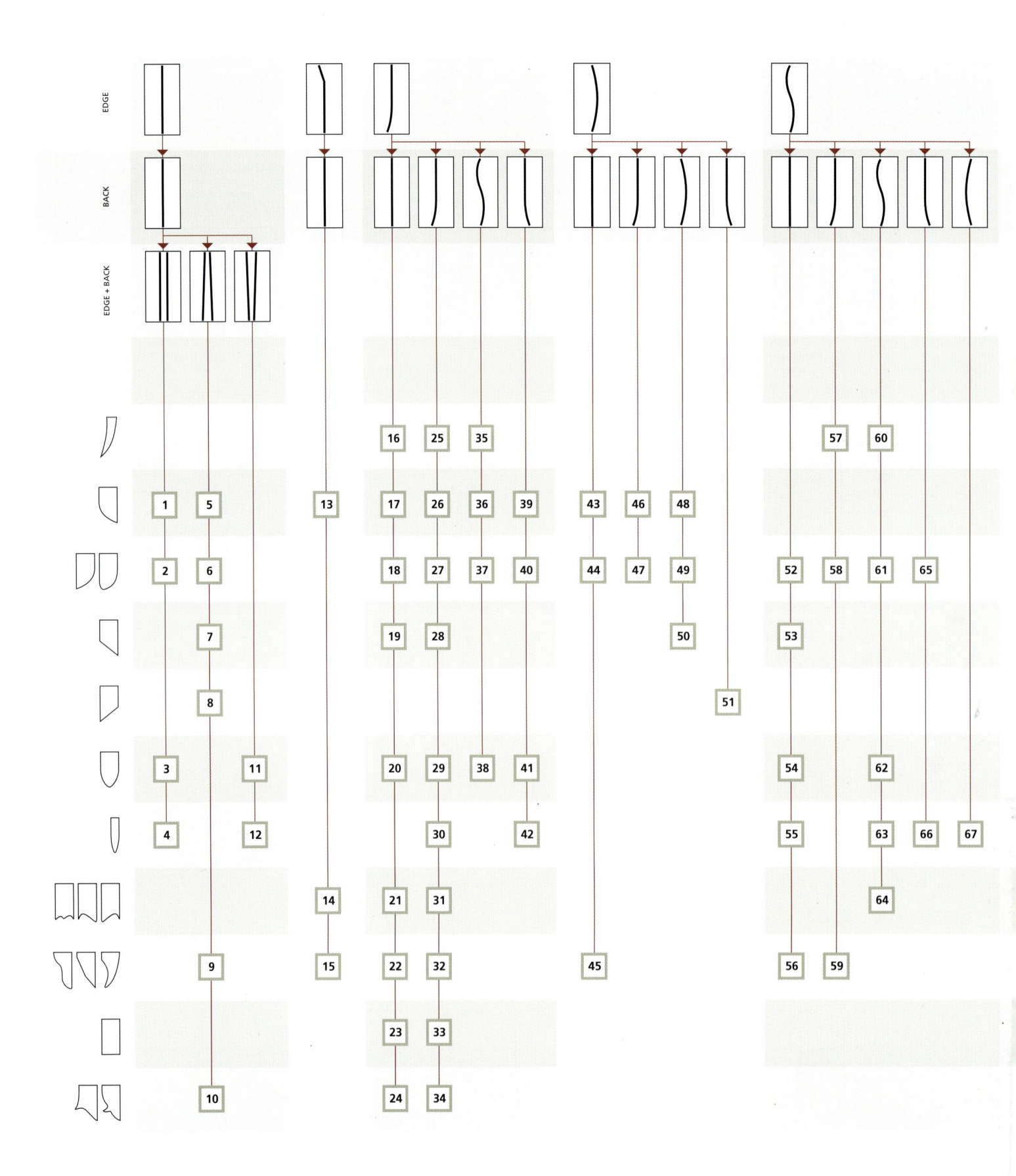

19
CLASSIFICATION

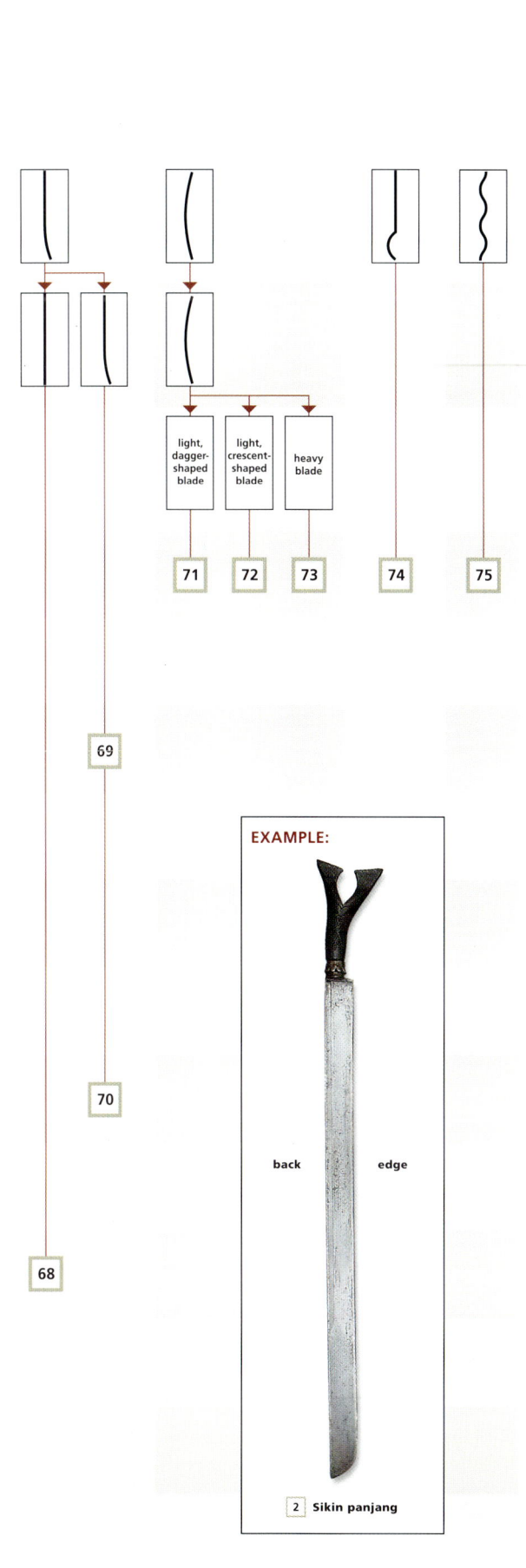

1
alamang
dua lalan
klewang
labo topang
pedang
penai

2
kawali
lopah petawaran
luju alang
raut
rawit pengukir
sakin
si euli
sikin lapan sagu
sikin panjang
swords of Tanimbar
swords of the Timor group

3
andar andar
badek
kaso
luju alang
parang panjang
pedang
pedang lurus

4
lading terus
palitai

5
candong
co jang
golok
hemola
kampilan
katungung
klewang
klewang tebal hujong
lopu
parang
swords of Tanimbar
swords of the Timor group
todo

6
amanremu
klewang
labo
ladingin
luju alas
parang
si euli

7
baradi
belida
klewang
luwuk
opi
parang lading
peda
stick sword
sumara
swords of Tanimbar
swords of the Timor group

8
amanremu
klewang

9
kampilan

10
kampilan
tampelan

11
chenangkas
pedang
sekin
swords of Tanimbar

12
luju celiko
taji

13
buko
parang latok

14
pandat

15
parang patah

16
klewang

17
kabeala
klewang

18
golok
parang
piso lampakan
piso ni datu
rawit
sikim gajah
tapak kudak
thinin

19
klewang
peda
swords of Tanimbar

20
badek

21
pandat
parang

22
mandau
parang
piso belati

23
parang lading

24
klewang

25
golok
parang
pedang
pedang bentok
pedang jenawi
pedang lurus
podang
shamshir
sikin pasangan
swords of the Timor group

26
chunderik
klewang
ladieng
matana knife
pakayun
parang
parang pedang
pedang
pedang bentok
piso raout
sikin pasangan

27
klewang
parang nabur
pedang
pedang bentok
piso raout
podang
to sangto

28
jimpul
klewang
labo balange
matana knife
moso
pakayun
parang
rugi

29
golok
pedang bentok

30
parang
pedang bentok

31
klewang tebol hujong
parang rantai

32
piso
tao

33
klewang
parang lading
parang upacara

34
klewang

35
pedang pemanchong

36
parang
tarah baju

37
labo bale bale
parang

38
barong
parang

39
klewang
lopu

40
badek
bodik
parang

41
arit
badek
barong
bayu
dohong
parang
parang panjang
piso
sadop

42
lading terus
tumbok lada

43
bangkung

44
badek
bendo
wedung

45
swords of Tanimbar

46
stick sword

47
cayul
golok
parang upacara

48
chandong
jimpul
langgai tinggang
swords of the Timor group

49
dukn
niabor
parang nabur
parang pedang
pedang bentok
pedang chembul
siraui

50
pakayun
parang

51
barong
thinin

52
gadoobang
jono
kalasan
kawali
pelaju
piso halasan
raut
rawit
rawit pengukir
surik
taka
wedung

53
bugis
swords of Tanimbar

54
andar andar

55
andar andar

56
piso sanalenggam

57
gadoobang
golok
pedang bentok

58
kalasan
piso eccat
piso gading

59
golok

60
arit lanchar
caluk lapar
golok
larbido
mentawa
raja dumpak
sepa
sewar
wedung

61
larbango
mentok
ronkepet
sekin
sewar
tigar

62
golok
keris
keris majapahit
keris panjang
kudi tranchang
pedang
sundang

63
rontegari

64
kudi

65
badek
bodik
klewang
peurawot
rencong
sewar
tumbok lada

66
pedang

67
arit

68
balato

69
chunderik
cok jang
corik
klewang
rudus
tarah baju

70
badek
gajang
sewar
tumbok lada

71
beladau
korambi
lawi ayam

72
arit
celurit
karis

73
arit
caluk
kudi
mundo
parang bengkok
parang ginah
ruding lengon
sadeueb
telabuna

74
balato
gari
si euli

75
keris
keris majapahit
sundang

EXAMPLE:

back — edge

2 Sikin panjang

ALPHABETICAL SURVEY

K. Podang hilt
Sumatra, Batak. RMV 3600-903. Acq. in 1959 through the Ethnografisch Museum van de Koninklijke Militaire Academie, Breda.

MANNER OF LISTING

HEAD NAME
The weapons in this book are described alphabetically, arranged according to their most frequent name: the head name.

Example:

HULU PEUSANGAN
[OELEE PEUSANGAN]

SUMATRA, ACEH, PEUSANGAN
A hilt that broadens at the top, as used with the *sikin panjang* or with the *pedang*. The hilt is almost straight and ends in two flattened protrusions, at the tip broad enough to almost touch each other. This form is characteristic of Peusangan in north Aceh.
(KREEMER; VOLZ 1912)

146. Hulu peusangan
North Sumatra. *Sikin pasangan* hilt. L. 15 cm.

SYNONYMS
All the synonyms and the names with different spellings, as found in the sources, follow the head name, between square brackets, where they are listed alphabetically. (All synonyms are included in the book as an entry too, with a cross-reference to the head name).

ORIGIN
The provenance follows the name. Firstly the island in question is named and, if known, the region or place (and the people) where the weapon was in use. If the provenance was either unknown or multiple, it is omitted.

DESCRIPTION
Detailed description of the weapon.

LEGENDS TO THE PLATES
Unless indicated otherwise, the indicated length for weapons is without scabbards (blade plus hilt).

LITERATURE
Sources from which the information provided was mainly drawn.

1. Alamang
South Sulawesi. RMV 1926-747. Acq. in 1916 through the Bataviaasch Genootschap van Kunsten en Wetenschappen, Batavia (Jakarta). L. 70.5 cm.

A a

ALPHABETICAL SURVEY

AGANG
[NGGILING]

WESTERN FLORES, MANGGARAI REGION
An oval or round shield, made of buffalo hide.
(DRAEGER)

AKAR BAHAR
[AKAR MAHAR, GABHA, KABHA, KARBAR, KERBATOE, KERBATU]

A species of coral consisting of a hard, often almost black, material used to make *keris* and *rencong* hilts. The term *akar bahar* is derived from the Arabic word *bahr*, meaning: 'sea'. The following types of this coral are found:
(a) *akar bahar (Plexaura)*, the most popular kind;
(b) *akar bahar belusop* has a rasp-like, rough surface and no branches. It is almost only used for making brooches;
(c) coloured or white *akar bahar punyuan (Antipathes sp.)* has fine branches and is mainly used for bracelets;
(d) white *akar bahar (Plexurella (Gorgonia) dichotoma)*, a raw material used for adornments.

In northern Sumatra the following words are used when referring to *akar bahar*: *kabha* or *gabha* (Aceh), *kerbatu* or *karbar* (Gayo) and *akar mahar* (Alas).
(JUYNBOLL 1910; KREEMER)

2. **Akar bahar**
North Sumatra. *Rencong* hilt.

AKAR MAHAR
See *akar bahar*

3. **Alamang hilt**
Sulawesi.

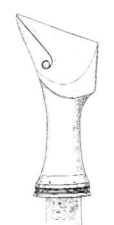

4. **Alamang hilt**
Sulawesi.

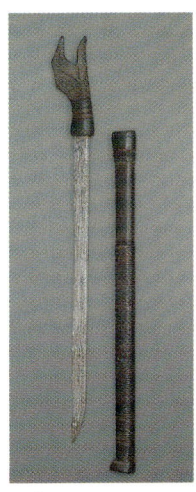

5. **Alamang**
Sulawesi, Toraja. L. 65.5 cm.

6. **Alamang**
South Sulawesi. RMV 1926-747. Acq. in 1916 through the Bataviaasch Genootschap van Kunsten en Wetenschappen, Batavia (Jakarta). L. 70.5 cm.

7. **Alamang**
L. 67.5 cm.

ALAMANG
[SONRI]

SULAWESI, JAVA
A Buginese sword with a flattened, heavy and deeply indented hilt. The blade's edge and back are straight and parallel. The back curves towards the edge at the point. The *alamang* has a straight, plain scabbard with sometimes a small foot. The scabbard's mouth is somewhat broadened. *Sonri* is the Makassar term for this sword.
(FISCHER 1914; MATTHES 1874; SCHRÖDER)

ALASAN
See *kalasan*

ALI-ALI
[BEMBAWANG]

SUMATRA, ACEH
A catapult or a sling, made of leather or plaited rattan.
(GARDNER 1936; KREEMER)

AMANREMOE
See *amanremu*

AMANREMU
[AMANREMOE, AMAREMOE, MEREMOE, MERMO, MERMOE, SAMAREMOE, SEMAREMOE]

NORTH SUMATRA
A long sword. The blade's edge and back are straight. It broadens towards an oblique point and has a slight curve from the edge to the back. The sides are smoothly forged. As the centre of gravity is situated at the point, one can deliver a heavy blow. The halves of the scabbard are held together with small metal or plaited rattan strips. Some scabbards are indented on the upper side.

The *amanremu* has the following types of hilts (sometimes decorated with a chased or embossed metal band):
(a) the *hulu babah buya* (Gayo: *hulu serampang*; Alas: *sukul ngango*) meaning: 'as the mouth of a crocodile';
(b) the *oelee iku mie* (Gayo: *hulu simpul*; Alas: *sukul simpul*) meaning: 'as the intertwined tails of cats';
(c) the *sukul jering* which strongly resembles (b).

The *amanremu* can be found in the Gayo, Alas and Batak regions (western Karo area, Pakpak area, Toba Batak) but is a rarity in the Laut Tawar region. The *sukul jering* is characteristic of the south-western Karo and Pakpak areas. It can also be found in the Toba and Alas regions and in Gayo Luos. The *hulu babah buya* we find in the Alas region, Gayo Luos and western Karo.
(JASPER 1930; KREEMER; VOLZ 1909, 1912)

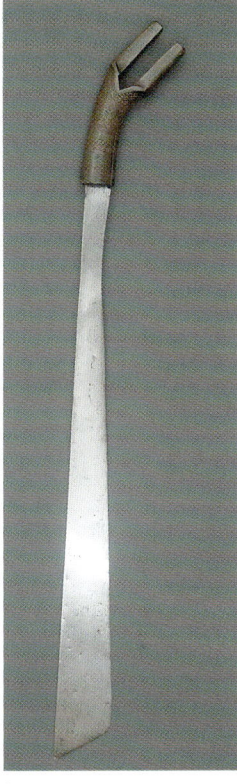

8. **Amanremu**
Sumatra, Aceh. Hilt: *hulu babah buya*. L. 66.5 cm.

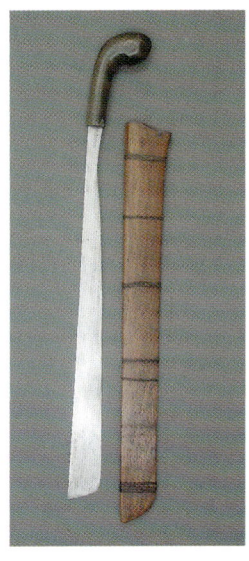

9. **Amanremu**
Sumatra, Aceh. Hilt: *hulu iku mie*. L. 58 cm.

AMAREMOE
See *amanremu*

AMBALANG
SUMATRA, TOBA BATAK
A type of sling.
(STONE)

AMPANG AMPANG
See *hampang hampang* and *peurise*

ANDAR
See *sikin panjang*

ANDAR ANDAR
[DJULUNG DJULUNG, JULUNG JULUNG]
SUMATRA, WEST KARO
A short sword with parallel, almost straight, back and edge coming together at the point. The back is sharpened from the point along c.1/3 of the blade. The hilt is forked (*sukul ngango*, *hulu babah buya* or *sukul nganga*) or thickened and curved at the top (*sukul jering*). It may be decorated by means of metal strips.
The scabbard's two parts are held together by metal or plaited rattan strips. Its upper half, or sometimes almost the entire scabbard, may have a broad metal covering. Its mouth widens and may have a decorated upper part called *sampir*.
(VOLZ 1909)

10. Andar andar
Sumatra, Batak. Hilt: *sukul ngangan*.
L. 52.5 cm.

11. Andar andar
Sumatra, Batak. Hilt: *sukul nganga*.
L. 64.5 cm.

12. Anggang gading
Kalimantan. Head of the *anggang totok*, with indication of the part of the beak used.

ANGGANG GADING
The largest hornbill found on the Malayan peninsula, Sumatra and Kalimantan is called *anggang totok* (*Rhinoplax vigil*). The ivory on its beak (*anggang gading*) consists of an excrescence measuring c.10 x 5 x 5 cm. Due to a cavity its thickest part measures c.1.5 cm. This semi-transparent, horn-like material is red on the outside and yellow on the inside, enabling the production of beautiful two-coloured objects such as beads, buttons, ear-rings and amulets. Such ear-rings (*sabau*) should only be worn by men who have participated in a head-hunting expedition. Moreover, buckles for the belts of *mandaus* are extremely finely carved with *aso* motifs: stylised dragons.
(VAN HEYST)

ANGGANG TOTOK
[TEBOEN, TEBUN]
The hornbill *anggang totok* which the inhabitants of the Apokayan region of Kalimantan, the Kayan and the Kenyah refer to as *tebun*.

13. Anggang gading
Central Kalimantan, Apo Kayan. The buckle of this *mandau* girdle is made of *anggang gading*.

ARECA NIBUNG
Wood of the *nibung* palm (*Areca nibung*) used for making lance-shafts.
(FISCHER 1909)

ARIT
SUMATRA, JAVA, MADURA, BALI
A sickle, tool-*cum*-weapon. The *arit* has many forms, depending on its use and place of origin. The blade usually has a more or less crescent shape. As a fighting weapon, it is carried in one hand, or in both, and sometimes in combination with a *piso*. See also *sadeueb*.
(DRAEGER: GARDNER 1936; KRUIJT; RAFFLES 1817)

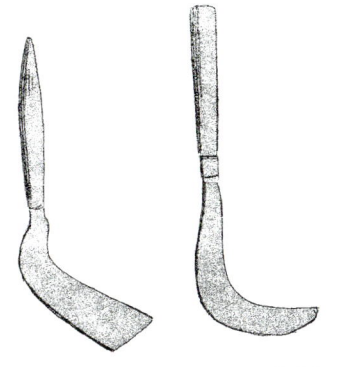

14. **Arit**
Java.

ARIT BIASA
MADURA
A sickle with a slightly curved blade.
(DRAEGER)

ARIT LANCHAR
MADURA
A sickle with an S-shaped blade.
(DRAEGER)

ARIT MENGOBED
BALI
A knife used for cutting grass and as a weapon. Its edge often has serrated teeth.
(DE KAT ANGELINO)

ASO MOTIF
A decorative motif in the form of a dog or dragon with a wide open mouth, full of teeth. The *aso* motif is often found on hilts and scabbards of *mandau*s.
(COPPENS; SELLATO)

AWOLA
[OLA, POESSOE POESSOE, PUSSU PUSSU, SAROBA]
SULAWESI
A spear, entirely made of bamboo or *pinang* wood. Known as *awola* (Buginese) and *saroba*, *ola*, *pussu pussu* (Makassar).
(MATTHES 1874, 1885; SCHRÖDER)

17. **Awola**
Sulawesi.

15. **Aso motif on a mandau hilt**
Kalimantan, Kayan.

16. **Aso motif on a mandau hilt**
Kalimantan, Kayan.

18. **Baru oroba**
Nias. RMV 2137-1. Acq. in 1927 from C.H.A. Groenevelt.
L. 60cm, W. 57 cm.

b

ALPHABETICAL SURVEY

BADA
See *kampilan*

BADANUMOGANDI
See *kampilan*

BADE
See *badek*

BADE GAGANG BEUSI
See *kudi*

BADEE
See *badek*

BADEH
SUMATRA, ACEH
A knife with a slightly curved blade which, prior to 1892, according to Snouck Hurgronje, had replaced the *rencong* amongst 'the men of religion'.
(SNOUCK HURGRONJE 1892)

BADEK
[BADE, BADEE, BADI, BADI BADI, BADIK, BADIK BADIK, BADIT]

SUMATRA, JAVA, MADURA, SULAWESI
A knife with a large variety of shapes and sizes, found all over the archipelago. The blade has a straight or slightly convex back running up to the point, or otherwise bends towards the edge at the point. The edge may be straight, but also slightly or strongly convex, or somewhat S-shaped. Sometimes golden figures are encrusted on the blade. These are called *jeko* (meaning: 'unjust'), on Sulawesi. Just above the blade, the hilt has an angle of 45-90°. The scabbard may have a small foot at the bottom. Its upper side can end straightly, but can also have a broader part made in one piece with the remaining part of the scabbard, or a small upper part made of a separate piece of wood.
The *badek* is carried either on the left or the right side of the body. The hilt's end points backwards in both cases. It is mainly used for stabbing, but also for slashing.
(DIELES; DRAEGER; EGERTON OF TATTON; FISCHER 1909, 1914; GARDNER 1936; GRUBAUER; KREEMER; PARAVICINI; RAFFLES 1817; STONE; VOSKUIL 1921)

BADI
See *badek*

BADI BADI
See *badek*

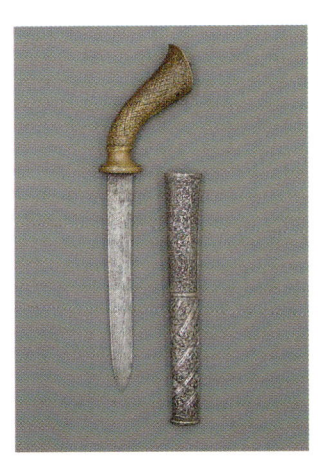

BADI GAGANG BEUSI
See *kudi*

BADI GOEROE
See *kawali*

BADI GURU
See *kawali*

BADIK
See *badek*

BADIK BADIK
See *badek*

BADIQ LOKTIGA
KALIMANTAN
A knife with a short pointed blade and a finely carved hilt.
(STONE)

BADIT
See *badek*

BADJOE-GOES
See *baju gus*

BADJOE-RANTE
See *baju rantai*

BADJOE-SAMBO
See *baju gus*

BAIENG
See *mandau*

BAJAU BELADAU
See *barong*

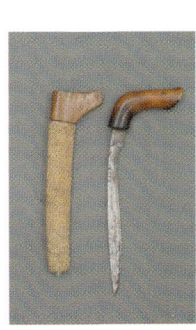

19. **Badek** Java (?). L. 34 cm.
20. **Badek** Sulawesi. L. 26 cm.
21. **Badek** Sulawesi (?). L. 18 cm.
22. **Badek** Sulawesi (?). L. 19 cm. 
23. **Badek** Java (?). L. 27 cm. 
24. **Badek** Sulawesi. L. 20 cm. 
25. **Badek** Java (?). L. 29.5 cm. 
26. **Badek** Sumatra, Palembang. L. 25 cm.

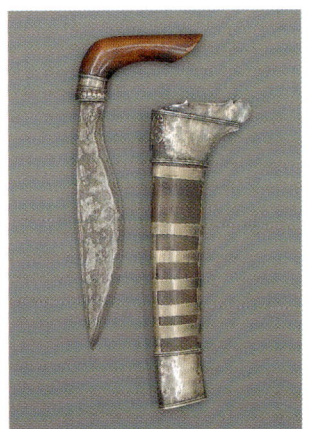

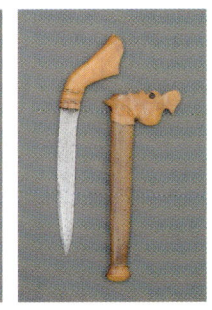

BAJU BELADAU
See *barong*

BAJU BESI
See *baju rantai*

BAJU EMPURAU
KALIMANTAN, SEA DAYAK
A kind of coat-of-mail made of thick bark and fish scales. The larger scales are attached by means of split rattan, the smaller ones by means of a strong cord. It has no sleeves or collar.
(ROTH 1896B)

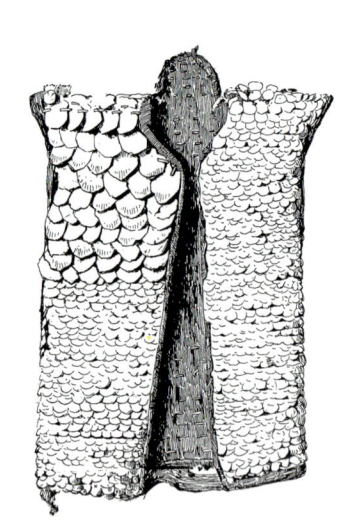

BAJU GUS
[BADJOE-GOES, BADJOE-SAMBO]

SUMATRA, ACEH
A long, padded war coat.
(KREEMER)

BAJU LAMINA
A cuirass resembling a coat-of-mail. It is made of brass rings to the front and back of which small rectangular pieces of brass are attached. This cuirass has no sleeves or collar.
(GARDNER 1936)

BAJU PA'BARANI
SULAWESI, SA'DAN TORAJA
A type of war dress.
(RODGERS)

BAJU RANTAI
[BADJOE-RANTE, BAJU BESI, BAJU RANTE, WADJOE-RANTE, WAJU RANTE]

SULAWESI
A coat-of-mail made of small iron rings. It has short sleeves and no collar. Known as *baju rante* (Makassar) or *waju rante* (Buginese).
(SCHRODER; MATTHES 1874, 1885; GARDNER 1936)

BAJU RANTE
See *baju rantai*

BAJU TILAM
See *klambi tayah*

BAKIN
KALIMANTAN, SEA DAYAK
A spear carried in the right hand.
(ROTH 1896B)

BAKONG
See *sadeueb*

BALADAU
SULAWESI, BOLAANG MONGONDOW
A knife made in Bolaang Mongondow in north Sulawesi.
(FOY)

BALAHOGO LEMA'A
NIAS
A head ornament for men to be worn in battle. It is made of two or three rings of plaited rattan and fibres of the *aren* palm (*Arenga sacchifera*). Sometimes it is decorated with pieces of flannel or cotton.
(FISCHER 1909)

BALASAN
SUMATRA, MINANGKABAU
A short blow-pipe made of a single joint of bamboo (*Bambusa wrayi* or *Bambusa longinodis*). Pieces of reed (*timbaru*) are attached to the rear of the darts which are wrapped around with raw cotton before being fired.
(DRAEGER; STONE; VOLZ 1909, 1912)

BALATO
[BALATOE, BALATU, BALATU SEBUA, BALLATU, FODA, GARI TELEGU, KLEWANG PUCHOK BERKAIT, ROSO SEBUA, TELAGOE]

NIAS
A sword with a large variety of blades, hilts and scabbards. Three types of blades can be distinguished, all broadening at the point:
(a) with an almost straight back and edge; the edge is rounded towards the back at the point;
(b) with an almost straight back and a straight or a somewhat concave edge where the back becomes S-shaped towards the edge;

27. Baju empurau
Kalimantan. Thick piece of bark, furnished with fish scales. L. 65 cm, W. 42 cm.

28. Baju lamina
South Sulawesi. RMV 522-1. Acq. in 1885 from dr. B.F. Matthes. L. 65 cm, W. 45.5 cm.

29. Baju rantai
Sulawesi.

30. Balato with amulets
Nias.

31. Balato
South Nias. L. 69.5 cm.

32. Balato
South Nias. L. 56.5 cm.

33. Balato
South Nias. L. 50 cm.

34. Balato
South Nias. L. 59.5 cm.

35. Balato
South Nias. L. 63 cm.

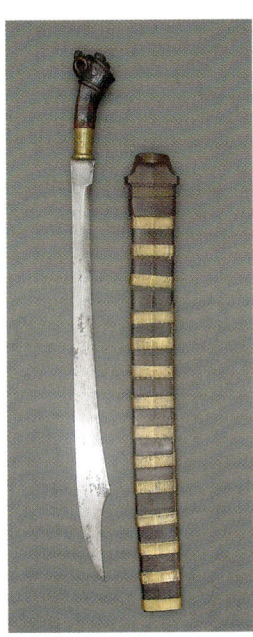

36. Balato
South Nias. L. 62.5 cm.

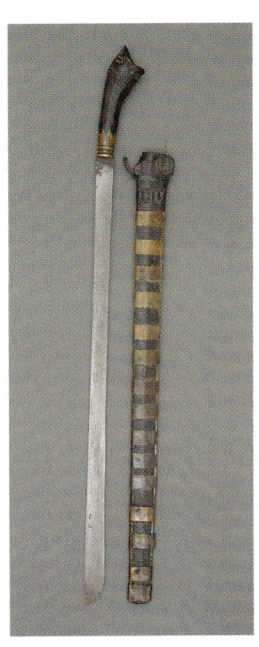

37. Balato
South Nias. L. 63 cm.

38. Balato
Nias. RMV 3600-1560. Acq. in 1959 through the Ethnografisch Museum van de Koninklijke Militaire Academie, Breda. L. 63.5 cm.

(c) with a slightly convex back, a slightly concave edge and a strongly concave segment near the point.

The hilts are very varied, but all can be reduced to an animal's head or mouth, often the *lasara*, executed in a plain stylised way or in a complex, richly decorated form. These hilts are mainly made of wood, but brass ones also occur. Wooden hilts have a broad metal (usually brass) ring broadening towards the blade.

The wooden scabbard's parts are often held together by numerous strips of metal or plaited rattan. Its narrow back and edge may be decorated with brass strips. A round plaited rattan basket is sometimes attached to the scabbard. This basket is decorated with large animal teeth and contains amulets serving as magical objects. Carved wooden amulet figures (*adu*) may be directly attached to the scabbard.

In north Nias fewer objects are attached to the basket's exterior in comparison with south Nias. There a rich variety of items is found such as:
(a) wooden or stone statuettes;
(b) fossil shark teeth, the local term for which may best be translated as 'thunder-stone' in analogy with belemnites (fossil dorsal bones of a species of cuttle-fish from the Jurassic period);
(c) crocodile teeth and tiger teeth (not with Nias as a habitat), reserved for the headmen and their next-of-kin, especially in the south;
(d) tiger claws, considered of great value especially in the south;
(e) rings allegedly found pierced through the noses of certain species of wild pigs, but in reality made of cast metal;
(f) all kinds of objects such as stones with a singular lustre or shape, rings, teeth (no pig teeth as pigs are continually threatened with being run through which could have an undesired influence on the owner), etc.

Baskets with anthropomorphic figures are scarce, those with animal teeth are more frequent. The latter type can be divided into two groups:
(a) with teeth covering the basket in such a way that their points come together on the outside of the basket;
(b) with teeth placed in such a way as to imitate the head of a *lasara*, a motif also seen on the hilt of these swords.

(BARBIER; FELDMAN 1990; FISCHER 1909; FORMAN; GARDNER 1936; DE LORM 1941, 1942; RODGERS; STONE; VOSKUIL)

BALATOE
See *balato*

BALATU
See *balato*

BALATU SEBUA
See *balato*

BALING-BALING
A set of iron bars connected by means of links to swing around and then hurl.
GARDNER 1936

BALLATU
See *balato*

BALOELANG
See *balulang*

BALOESE
See *baluse*

BALULANG
[BALOELANG]

SULAWESI, TORAJA

A rectangular metal shield. The upper and lower side are somewhat convex. The two other sides are somewhat concave.
(DRAEGER; LOEBER 1939)

BALUSE
[BALOESE]

NIAS

An oval, leaf-shaped wooden shield. Its lower side ends in a solid long round extension with a rounded end or sometimes with a slight point. The upper part has a number of shallow notches on both sides and tapers towards a short flat extension. From the top downwards there runs a raised rib interrupted in about the centre by an oval or round knob. On the inside of the *baluse*, the hilt and a space for the hand have been cut out. This shield is made of a single piece of wood often decorated or strengthened with horizontal cords of rattan serving to prevent it from splitting lengthways when struck by an unfriendly sword. Its surface is usually undecorated, but may have intricately carved images or painted motifs.
(DRAEGER; FELDMAN 1990; FISHER 1909; KOL; RODGERS)

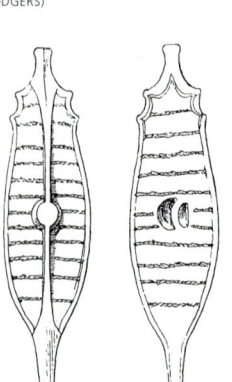

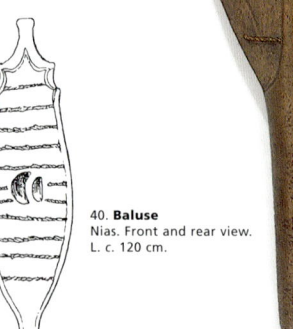

39. Baling baling

40. Baluse
Nias. Front and rear view. L. c. 120 cm.

41. Baluse
Nias. RMV 3600-1272. Acq. in 1959 through the Ethnografisch Museum van de Koninklijke Militaire Academie, Breda. L. 115.5 cm, W. 27 cm.

42. Baluse
North Nias. Warrior with a *baluse* (shield), a *toho* (spear) and a *balato* (sword).

BANDANG
JAVA

A spear with a normal point and a knob. A cord is attached to a small tassel on its lower side.
(STONE)

BANDOL
[BANDUL]

JAVA

A sword with a hook-shaped point.
(RAFFLES 1817A; STONE)

BANDRING
JAVA

A sling to throw stones with. According to Raffles, this weapon was used in 1812 against the British.
(RAFFLES 1817A; STONE)

BANDUL
See *bandol*

BANGKOENG
See *bangkung I*

BANGKUNG I
[BANGKOENG, BERANG, PARANG]

SULAWESI

A short, firm machete with a straight back and convex edge tapering at the point. The hilt is short and thick. *Bangkung* is the Buginese term for *berang* (*parang*).
(MATTHES 1874, 1885; SCHRÖDER)

BANGKUNG II
[RAUT RAUT]

A weapon from Sumatra.
(ROGERS)

BARA
See *kampilan*

BARADI
JAVA

A sword with a broad blade. Its edge and the back run almost parallel. At the end and at the hilt, the edge is longer than the back.
(JASPER AND PIRNGADIE 1930)

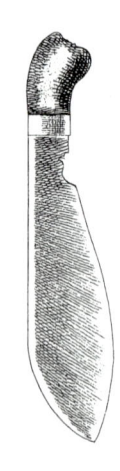

43. Bandring
Java.

44. Bangkung
Sulawesi.

45. Baradi
Java.

BAROE LEMA'A
See *baru lema'a*

BAROE SINALI
See *baru sinali*

BARONG
[BAJAU BELADAU, BAJU BELADAU, KLEWANG BELADAN, SULU KNIFE]

KALIMANTAN, SULAWESI, SULU

The *barong* is the national weapon of the Moros of Sulu, Mindanao (Philippines). It is also found in north Kalimantan and Sulawesi. This sword has a short, heavy blade. The edge and, in a lesser degree, the back are convex. They meet in a sharp point. Thanks to the blade's size, it is very effective in jungle combat or in places where space is limited. The blade usually has one sharp side. Sometimes the back is sharpened until about halfway. The hilt curves towards the end and often has two points. It may be plain, without any distinct points or projections. It may also end in beautifully carved projections, one of which curves towards the blade. This serves to prevent the sword from slipping out of the hand in combat. Hilts of the *barong* are made of wood, horn, ivory, silver, gold or a combination thereof. They often have a ring around the hilt's end, near the blade, or a sleeve extending halfway up the hilt. The hilts are related to those of the *sulu keris*. The *barong*s with beautifully carved hilts mainly serve as weapons of state, those with plainer hilts are used as battle-weapons. The *barong* is carried in a flat, wooden scabbard. Its opening widens and is often decorated with carving. A broader part, sometimes carved, often forms the scabbard's foot.
(CATO; DIELES; FORMAN; GARDNER 1936; RAFFLES 1817A; STONE)

46. **Barong** Kalimantan. L. 61 cm.

47. **Barong** Kalimantan.

48. **Baru lema'a** Nias. RMV 1002-137. Acq. in 1894 from Palmer van de Broek. L. 64 cm, W. 49 cm.

BARU LEMA'A
[BAROE LEMA'A]

NIAS

A sleeveless war jacket made of woven *ijuk* fibre. The back part ends at the top with a decoration: two round ornaments meant to protect the neck.
(FELDMAN 1990; FISCHER 1909)

BARU OROBA

NIAS

A leather cuirass with broad shoulder parts and large vertical folds. The back part ends at the top with a decoration of two round ornaments meant to protect the neck. This type of cuirass is sometimes made of iron. See also *g-orbana'a*.
(FELDMAN 1990; RODGERS)

49. **Baru oroba** Nias. RMV 2137-1. Acq. in 1927 from C.H.A. Groenevelt. L. 60cm, W. 57 cm.

50. **Baru oroba** Nias. Rear view of illus. 49.

BARU SINALI
[BAROE SINALI]

NORTH NIAS

A sleeveless war jacket, woven with plant fibres. The edges may be hemmed with textile.
(FISCHER 1909)

BASI AJE TADO
[BASI WAKO TADO, BASSI ADJE-TADO, BASSI WAKKO-TADO, BOELO-TADO, BULU TADO]

SULAWESI

A lance, the front part of which has a noose made of rope. *Basi aje tado* and *basi wako tado* are Buginese terms, *bulu tado* is Makassar.
This weapon has the following parts: *mata* (Buginese/Makassar): point; *tentengang* (Buginese/ Makassar): shaft; *cappakang* (Buginese), *sipi* (Makassar): slip-knot; *tulu tado* or *ula kulang* (Buginese) *otere tado* (Makassar): noose.
(MATTHES 1874, 1885; SCHRÖDER)

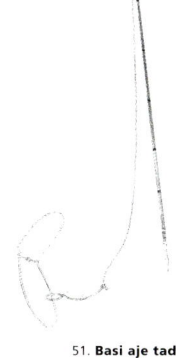

51. **Basi aje tado** Sulawesi.

52. **Basi baranga** Sulawesi.

BASI BARANGA
[BASSI-BARANGA, POKE-BANRANGANG]

SULAWESI

A spear with at the bottom of the shaft a 'flag' mainly consisting of horse's or goat's hair. This weapon is called *basi baranga* in Buginese and *poke banrangang* in Makassar. The names given to the parts of this spear are: *mata-basi* (Buginese), *mata-poke* (Makassar): point; *tentengang* or *pasorang* (Buginese/Makassar): shaft; *banraga* (Buginese), *banrangang* (Makassar): hair; *wanuwa* (Buginese), *bannowa* (Makassar): scabbard; (gold or copper) *pando* (Buginese/Makassar): mount.
(MATTHES 1874, 1885; SCHRÖDER)

BASI KALIYAWO
See *basi sanresang*

BASI KUMPA
See *basi sanresang*

BASI PAKA
[BASSI-PAKKA, POKE PANGKA]

SULAWESI

This elegant spear has a forked point. Known as *basi paka* (Buginese) or *poke pangka* (Makassar).
(MATTHES 1874, 1885; SCHRÖDER)

53. **Basi paka** Sulawesi.

BASI PARUNG
[BASSI-PAROENG, POKE-PAROENG, POKE PARUNG]

SULAWESI

A spear with an undulating point. Known as *basi parung* (Buginese) or *poke parung* (Makassar).
(MATTHES 1874, 1885; SCHRÖDER)

54. **Basi parung** Sulawesi.

BASI SANGKUNG
[BASSI-SANGKOENG, POKE-SANGKOENG, POKE SANGKUNG]

SULAWESI

A spear with a large point and a shaft tapering somewhat towards the base. It is referred to as *basi sangkung* (Buginese) and *poke sangkung* (Makassar).
(MATTHES 1874, 1885; SCHRÖDER)

BASI SANRESANG
[BASI KALIYAWO, BASI KUMPA, BASSI-KALIYAWO, BASSI-KOEMPA, BASSI-SANRESANG, POKE-TANJDJENGANG, POKE TANJENGANG]

55. **Basi sangkung** Sulawesi.

SULAWESI

A spear the point of which has a cover of red cloth. It is named *basi sanresang*, *basi kaliyawo* or *basi kumpa* (Buginese) and *poke tanjengang* (Makassar). Its cover is called *sepu* or *pasaru* (Buginese) and *sarombong* (Makassar).
(MATTHES 1874, 1885; SCHRÖDER)

BASI TAKANG
[BASSI-TAKKANG, POKE TAKANG, POKE-TAKKANG]

SULAWESI

A spear also used as a stick or staff. It is referred to as *basi takang* (Buginese) and *poke takang* (Makassar).
(MATTHES 1874, 1885; SCHRÖDER)

56. **Basi sanresang** Sulawesi.

BASI WAKO TADO
See *basi aje tado*

BASSI ADJE-TADO
See *basi aje tado*

BASSI-BARANGA
See *basi baranga*

57. **Basi takang** Sulawesi.

BASSI-KALIYAWO
See *basi sanresang*

BASSI-KOEMPA
See *basi sanresang*

BASSI-PAKKA
See *basi paka*

BASSI-PAROENG
See *basi parung*

BASSI-SANGKOENG
See *basi sangkung*

BASSI-SANRESANG
See *basi sanresang*

BASSI-TAKKANG
See *basi takang*

BASSI WAKKO-TADO
See *basi aje tado*

BATU RAJUT
A metal ball in a net, used as a projectile.
(GARDNER 1936)

58. **Batu rajut**

BAWAR
[KERODJON, KEROJON, MERHOEM, MERHUM, PERMATA]

SUMATRA, ACEH

An ancient type of knife accredited with mysterious power. The *sultan* of Aceh presented such knives to the *kedjuruns* and to some minor chiefs as insignia of their office, or to acknowledge their dignity.
(KREEMER)

BAYU
KALIMANTAN, DAYAK

A weapon related to the *mandau*. The scabbard and hilt are the same as those of the *mandau*, only the blade differs. Unlike the sides of the *mandau*, of which one is concavely and the other convexly shaped, the sides of the *bayu* are flat. Moreover, the blade is for the largest part double edged. Its broadest part is near the point. The edge is almost straight, the back is distinctly curved towards the edge, near the point.

The *bayu* consists of the following parts:
(a) *iki*: a hilt;
(b) *balut*: a ring made of *damar* (black resin) covering the blade's insertion in the hilt;
(c) *krowit*: a guard for the fingers;
(d) *kowal*: embossed lines on the blade (which the Kayan call *karan*);
(e) *lantak paku*: small brass pins placed in the blade;
(f) *tandup*: a triangular area on the scabbard's exterior (which the Kayan call belipap);
(g) *sakum*: a bone plug at the end of the scabbard;
(h) *kowit*: bindings of the scabbard;
(i) *jubor*: hair;
(j) *supei*: a belt.
(GARDNER 1936; SHELFORD; STINGL; STONE)

BEADAH
See *niabor*

BECHOE
See *bekhu*

BECHU
See *bekhu*

BEDER
JAVA

An arrow to be shot with a bow.
(RAFFLES 1817A)

59. **Beder** Java.

BEKHU
[BECHOE, BECHU]
SOUTH NIAS
A figurine representing a demon (*bekhu*). This motif is found on *lasara* sword hilts of the *balato*s of south Nias. It resembles a human being or animal lying on its stomach, and is located on the head or in the neck of the *lasara*. A small bar forms the connection between the *bekhu*'s open mouth and the *lasara*'s head. This bar could represent the *bekhu*'s tongue or the skin of the *lasara*'s neck. The *bekhu* may have clearly recognisable arms with hands, legs with feet, eyes and ears, but sometimes only in a stylised manner. The *bekhu*'s position may differ. Its arms can point forwards. Otherwise, one or both arms hold its head or the small bar connecting its head with the *lasara*.
(FISCHER 1916; DE LORM 1941A)

60. Bekhu on a balato hilt
South Nias. L. 15 cm.

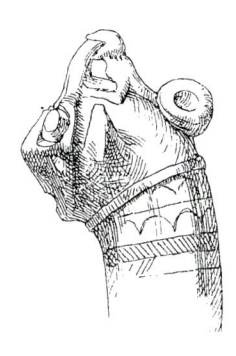

61. Bekhu on a balato hilt
South Nias.

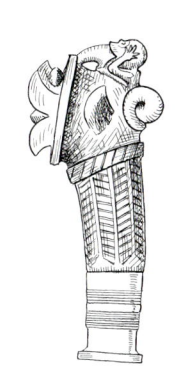

62. Bekhu on a balato hilt
Nias.

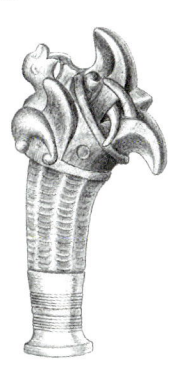

63. Bekhu on a balato hilt
Nias.

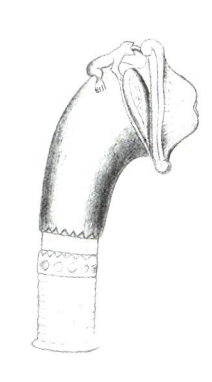

64. Bekhu on a balato hilt
Nias.

BELABANG
See *parang nabur*

BELADAH
See *parang nabur*

BELADAU
SUMATRA, MENTAWAI ISLANDS, RIAU ARCHIPELAGO
A short and curved knife. Its blade is broadest at the hilt and clearly curves towards a pointed tip. If this solid weapon has one sharp side its cutting edge is on the concave side. *Beladau*s with two sharp sides are also found. The blade often has a thickened or heightened ridge from the hilt on, along the blade's centre to the tip. The *beladau* is held underhand with the thumb near the blade and thrust upwards.
(DRAEGER; GARDNER 1936; HILL 1970)

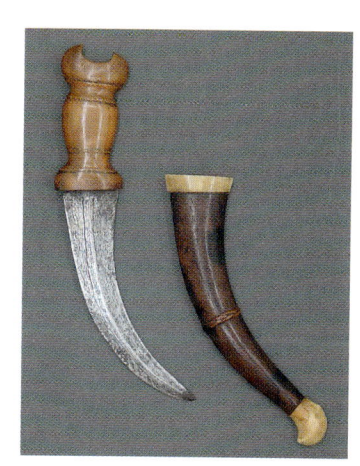

65. Beladau
L. 24 cm.

BELANGAH
KALIMANTAN, DAYAK
A blow-pipe with a decorated end and scabbard.
(STONE)

BELIDA
FLORES
A broadly tapering sword with an almost straight edge and back. The latter ends at the point in a curve towards the edge. The *belida* has a large beautifully elaborated hilt which can be held in both hands. The hilt's lower and upper part are separated by means of a decorated cross-piece.
(FOY)

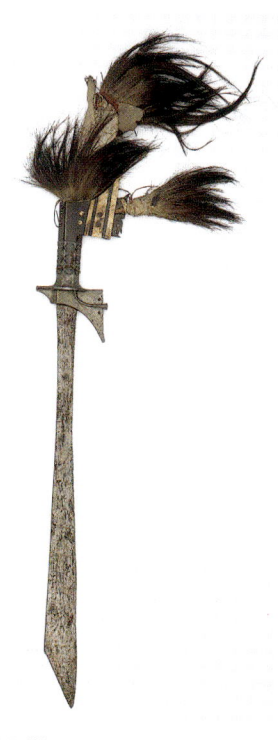

66. Belida
Solor. RMV 16-103. Acq. in 1861 from dr. S. Müller. L. 93 cm.

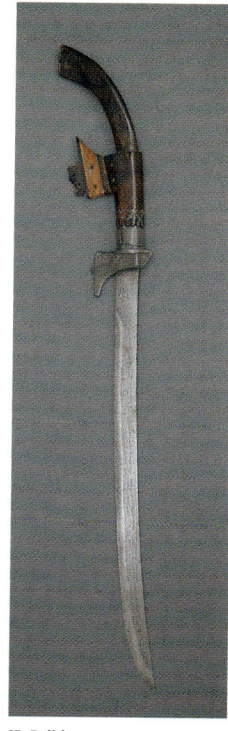

67. Belida
Flores. Unusually shaped blade. L. 82 cm.

68. Belida
Flores. RMV 798-3. Acq. in 1890 through the Koninklijk Nederlandsch Aardrijkskundig Genootschap. L. 79.5 cm.

69. Belida
Solor. Warrior carrying a sword (*belida*).

BELO
[BELO-LEONG]
FLORES
A weapon measuring between 65-150 cm about half of which consists of a leaf-shaped blade. In time of war the *belo* is called *belo leong*. See also the entry: stick sword.
(DRAEGER)

BELO-LEONG
See *belo*

BEMBAWANG
See *ali-ali*

BENDO
JAVA
A machete in the shape of an axe, mainly used for harvesting the *aren* palm. The short, broad and heavy blade serves this purpose excellently. The *bendo* often has a finely carved hilt made of wood or bone.
(PARAVICINI)

70. Bendo
West Java. L. 31 cm.

71. Bendo
Java, Bogor. L. 25.5 cm (including scabbard).

BERANDAL
KALIMANTAN
A heavy, two-handed sword, mainly found on Kalimantan.
(HILL 1970)

BERANG
See *bangkung* and *parang*

BEREGOE
See *beregu*

BEREGU
[BEREGOE]
SUMATRA, ACEH
A bugle made of buffalo horn used to call the warriors to battle.
(JACOBS)

BESI DUA
JAVA, BALI
A spear with two points.
(GARDNER 1936)

72. Beregu
Sumatra, Aceh. L. 39.5 cm.

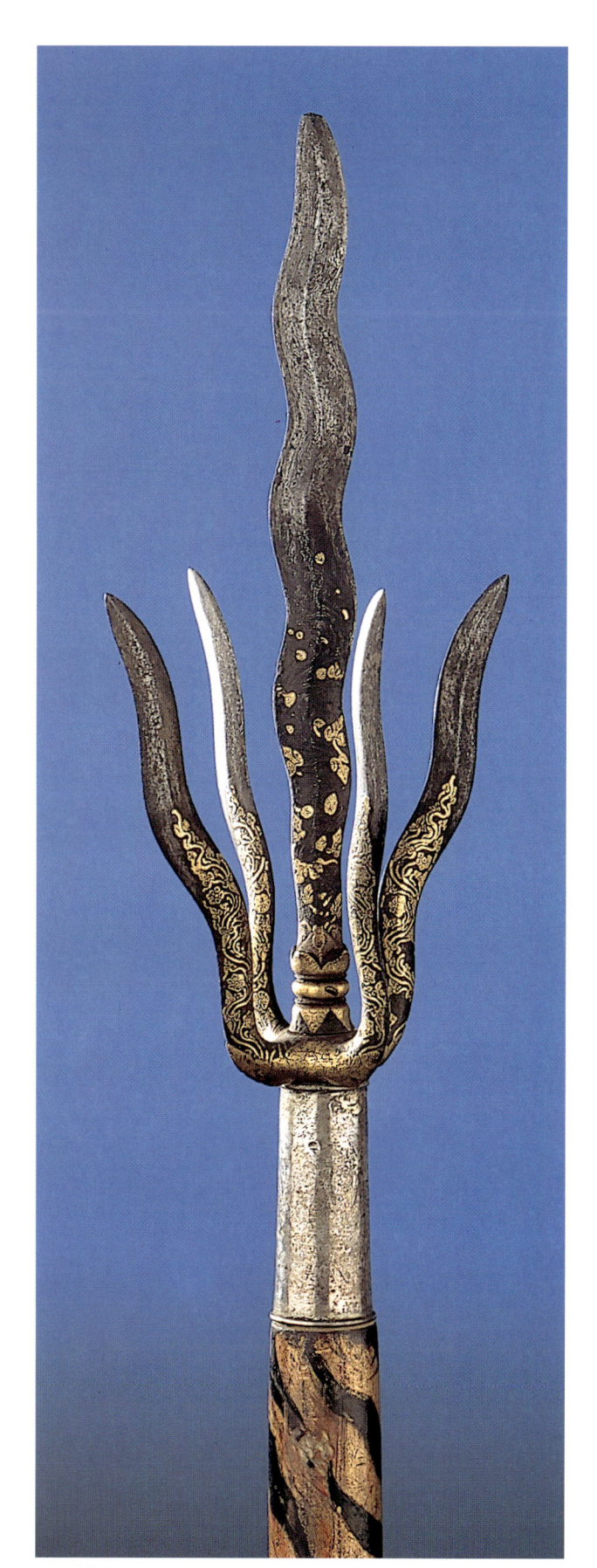

ALPHABETICAL SURVEY

BESI LIMA
JAVA, BALI
A spear with five points, the middle one being the longest.
(GARDNER 1936)

BESI TIGA
See *trisula*

BESSI TJABANG
See *cabang*

BEULANGKAH
See *sikin pasangan*

BLASUNG
KALIMANTAN, KAYAN
A large, flat shell attached to the front of a war jacket serving as protection.
(ROTH 1896B)

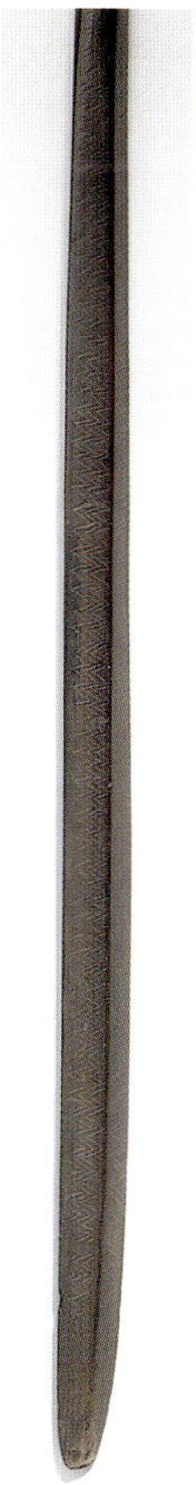

74. **Besi lima**
L. 25 cm.

BLUDGEON
[CLUB]
SULAWESI, TIMOR, ENGGANO, MALUKU, RIAU AND THE LINGGA ARCHIPELAGO

Clubs were made in various shapes and sizes, often with sharp protrusions. On Enggano the women sometimes used long bludgeons to destroy the shields of their opponents in combat.
(OUDEMANS; DRAEGER)

BODIK
JAVA
A short, broad sword with a slightly convex back and a somewhat S-shaped edge.
(JASPER AND PIRNGADIE 1930)

BOEGIS
See *bugis*

BOEKBOEK
See *bukbuk*

75. **Bludgeon**
Enggano. RMV 886-19. Acq. in 1892 from O.L. Helfrich. L. 137 cm.

BOEK-BOENG
See *bukbuk*

BOELOE-TADO
See *basi aje tado*

BOELOE SEWA
See *bulu sewu*

BOLADO
See *tumbok lada*

BOLANGKAH
See *sikin pasangan*

BOLONGKIDANG
SULAWESI
A type of knife made specifically in Bolaang Mongondow, north Sulawesi.
(FOY)

76. **Bodik**
Java.

73. **Besi lima**
Java. RMV 1438-1. Acq. in 1904. L. 65 cm.

BUGIS
[BOEGIS]

JAVA

A sword with a heavy, broad blade. Its back is almost straight. Its edge is S-shaped. The blade's narrowest part is c.1/3 of the way along the blade, from the hilt. The broadest part is not far from the point. The back ends in an oblique angle towards the edge.
(JASPER AND PIRNGADIE 1930)

BUKBUK
[BOEKBOEK, BOEK-BOENG, BUK BUNG]

MENTAWAI ISLANDS

A quiver with lid. Around the quiver plaited rings of rattan (in a fish-bone pattern) are placed. The bukbuk has a strap so that it can be carried over the shoulder.
(FISCHER 1909)

BUK BUNG
See *bukbuk*

BUKO
[BUKU]

KALIMANTAN, LAND DAYAK

This *parang* of the Land Dayak is smaller than the *parang latok*. Its blade is forged out of a square piece of iron, angled near the hilt. The longer part tapers into a broad blade. From the point to the bend the blade measures c.45 cm. The part from the bend to the hilt measures c.8 cm and has a rectangular cross-section. The blade's broadest part measures 4-5 cm.
The back is somewhat shorter than the edge and has a strong curve towards the edge. The back's thickness causes the blade's cross-section to be wedge-shaped in the centre.
The upper part of the wooden hilt is formed by a concave knob facing the front. Its top is convex and has diagonal grooves. The knob's sides are flattened. Plaited rattan may be wrapped around the hilt, or it can be mounted with silver.
The Bennah people of the main waters of the Sarawak River use a *buko* with a smaller hilt. The Betah of the Quop river (a branch of the Sarawak River) use a larger hilt and the Sempok an even larger one. The Pinyawa of the Samarahan river do not decorate hilts at all, but flatten the upper part until it has a triangular shape when looking from the side.
The scabbard is straight and covers the blade up to the bend. Its mouth is carved in deep relief. At its end, holes may have been made in which plugs are placed in order to secure tassels of hair. Both parts of the scabbard are held together by means of small strips of plaited rattan, *burad*. Scabbards with five small strips are called *burad patung*; those with seven small strips: *burad kiping*; those with nine small strips: *burad tipiris*; scabbards with eleven, thirteen, thirteen, or nineteen small strips: *burad brad bodad*. The belt, *taris*, is made of the veins of the bracts of a palm-tree.
(GARDNER 1936; SELLATO; SHELFORD)

BUKU
See *buko*

BULU AJAM
See *bulu ayam*

77. **Bugis**
Java.

BULU AYAM
[BULU AJAM]

MADURA

A sickle-shaped weapon (*bulu ayam*, meaning: 'chicken feather'). In general, it is longer and more curved at the hilt in comparison with the *arit* of Java. Two such weapons can be used in combat at the same time. A combination with the *pisau* is also possible.
(DRAEGER)

BULU SEWU
[BOELOE SEWA]

NIAS

A sword with a straight-backed blade which tapers towards the point. The edge is straight with at the front a convex curve towards the point, and is sharpened along almost the entire length.
(FISCHER 1909)

BULU TADO
See *basi aje tado*

BUMBAWE TEFAO
See *bum'bere tofao*

BUM'BERE TOFAO
[BUMBAWE TEFAO]

NIAS

A facial decoration to which various projections are attached. It is worn as a fanciful moustache forming part of a battle-kit meant to frighten the enemy.
(FELDMAN 1990; RODGERS; STONE)

BUNO
See *randu*

BUNO HAMPAK
See *randu*

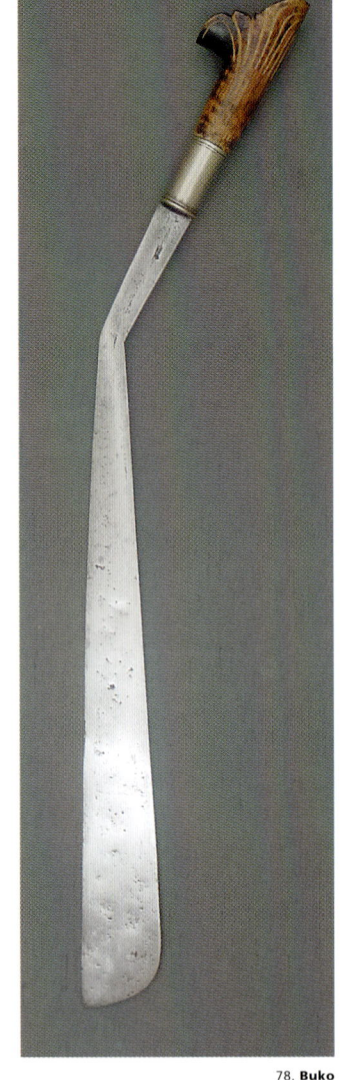

78. **Buko**
Kalimantan. L. 58 cm.

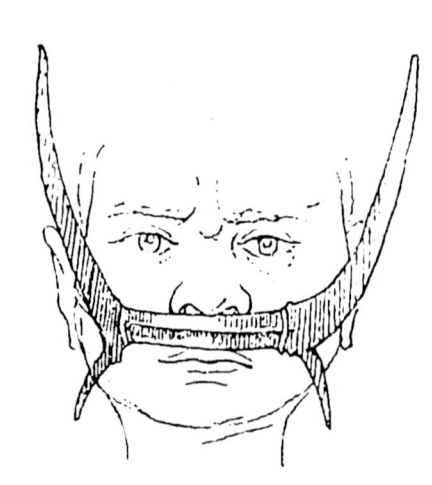

79. **Bum'bere tofao**
Nias.

37
ALPHABETICAL SURVEY

80. **Balato**
Nias. RMV 3600-1560.
Acq. in 1959 through the
Ethnografisch Museum
van de Koninklijke
Militaire Academie,
Breda. L. 63.5 cm.

81. Double knife
Sumatra, Karo Batak. RMV 2636-7. Acq. in 1947.
L. 28/21 cm.

CD
cd
ALPHABETICAL SURVEY

CABANG
[BESSI TJABANG, TITJIO, TJABANG]

JAVA, MADURA, BALI, SUNDA ISLES, SULAWESI, MALUKU
A short metal pole (measuring from 25-50 cm) with two protrusions appearing from the shaft at the place where the hilt ends. The *cabang* is perhaps derived from the Indian *trisula*, trident, and dates from the pre-Majapahit period (perhaps Srivijaya). It was originally used as a means of defence or to ward off blows. The two protrusions are usually at a right angle, curving towards the point. On Sulawesi, one of the protrusions curves in the opposite direction. At Batumerah in Maluku this weapon is called *besi cabang* (iron branch).
(DRAEGER)

CALUK
[TJALUK]

JAVA
A short curved knife of Muslim provenance. The blade is strongly curved. Its concave side is sharp, making it difficult to block a blow without injuring oneself. The hilt has a hand protector. The *caluk* has its greatest value in combat at close quarters. Due to its size (c.2/3 of the average *arit*), it can easily be concealed in loose clothes.
(DRAEGER)

CALUK LAPAR
[TJALUK LAPAR]

A fancifully-shaped knife of Muslim provenance. The word *lapar* (meaning: 'hungry') refers to the fact that the knife wishes to be used.
(DRAEGER)

CAMPILAN
See *kampilan*

CANDONG
See *rudus*

CAYUL

JAVA
A machete, mainly used as an agricultural tool. The *cayul* has a short, broad blade with a convex sharp side and a slightly concave back. Its point is rounded, its hilt quite long.
(RAFFLES 1817)

CELURIT
[TJELURIT]

MADURA
A crescent-shaped weapon with a blade which curves more than half a circle. Possibly used as an agricultural tool in the Banjuwangi region on east Java and then conveyed to Madura.
(DRAEGER)

CEREMONIAL AXE

BALI
Various types of axes are used on Bali during Hindu rituals and ceremonies. Their long, round hilts become somewhat thinner to one or both ends. The blade usually has a fanciful shape and encrusted motifs.

83. Ceremonial axe
Bali. L. 42 cm.

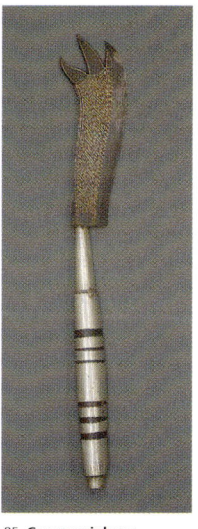

84. Ceremonial axe
Bali. L. 39.5 cm.

85. Ceremonial axe
Bali. L. 51 cm.

CHACRA
[CHAKRA]

JAVA
A mythical arrow used by gods. The upper part is wheel-shaped with protruding spokes.
(RAFFLES 1817)

CHAKERA

JAVA
A toothed discus mentioned in historical literature and depicted on Indo-Javanese monuments.
(GARDNER 1936)

86. Chacra
Java.

CHAKRA
See *chacra*

CHANDAK
A spear with a short, firm shaft.
(GARDNER 1936)

CHANDONG

KALIMANTAN, DAYAK
A machete used by the Dayak of the Baram river. The blade's edge is convex. The back is concave and curves towards the edge at the point. The centre of gravity lies near to the tip. The wooden scabbard's two parts are held together by means of rattan strips. The scabbard may be finely decorated, for instance with open-work carvings of floral motifs.
(ROTH 1896A)

CHELANA
A pair of trousers, forming part of a battle suit and entirely covered with buttons.
(GARDNER 1936)

CHEMETI

JAVA, SUMATRA, NUSA TENGGARA
A whip made in the Ponorogo region of east Central Java made of buffalo leather, human hair or a metal chain. Its hilt is usually made of hard leather or hardwood with leather wrapped around it. The total length varies between 90-150 cm. It is used in a mar-

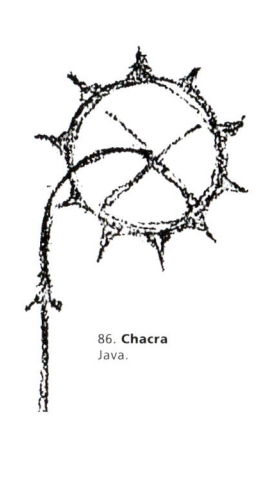

82. Cayul
Java.

tial art named *cambuk*.
The Batak on Sumatra sometimes use the *chemeti*, too. When found in Nusa Tenggara, it measures between 150-180 cm and is usually made of the main veins of palm-stalks tied together by means of strips of buffalo leather with interspaces along its length. More stalks but with a smaller diameter are used towards the point, making the tip thin but very flexible.
(DRAEGER)

CHENANGKAS
A straight variant of the *pedang I*. Its blade has the same width over the entire length. Its hilt is usually made of copper. However, sometimes hilts of iron, silver or embossed bronze occur.
(GARDNER 1936)

CHOKUMER
A sceptre or club, covered with nails. The *chokumer* is known from historical literature only.
(GARDNER 1936)

CHUNDERIK
[CHUNDRIK, CONDRE, CUNDRIK, TJOENDRIK, TJONDRE]

The name given to weapons of widely differing forms. For instance: a long sabre, a *klewang* or a small dagger the blade and hilt of which are forged from one piece and thus related to the so-called *keris majapahit*. The *chunderik* may sometimes serve as the point of a spear.
(GARDNER 1936; HAMEL; HAZEU; RAFFLES 1817A; SNOUCK HURGRONJE 1892; VERMEIREN 1987)

CHUNDRIK
See *chunderik*

CHURA SI-MANJAKINI
A *chunderik* of the dynasty of Indonesian *sultan*s.
(GARDNER 1936)

CINANGKE
[TJINANGKE]

CENTRAL SUMATRA
A variant of the *klewang*. The scabbard is usually made of *surian* wood. Its halves are held together by plaited rattan bands. See also *rudus*.
(ROGERS, VOSKUIL 1909)

CIO
[TJIO]

JAVA, MADURA
A spear with a diamond-shaped point.
(DRAEGER)

CLUB
See bludgeon

CO JANG
[GLIWANG TJOT JANG, KLEWANG COK JANG, KLEWANG TJOK JANG, KLEWANG TJOT JANG, TJO JANG, TJOK IJANG, TJOK JANG, TJOK YANG, TJOQ JANG, TJOT JANG]

SUMATRA, ACEH
A sword named after the *kampong* (village) from which this type allegedly originates. Its blade sometimes shows beautiful lines, produced by the forging together of iron and steel. The blade's back is almost straight, the edge slightly concave. The back runs to the edge in a curve. The sides are flat, without grooves or ribs. The hilt is of the *hulu tapa guda* type. The *co jang* is worn without a scabbard. As protection the blade may be wound in palm-leaf or goat's skin.
(KREEMER; VAN LANGEN; NIEUWENHUIZEN; ROGERS; SNOUCK HURGRONJE 1904)

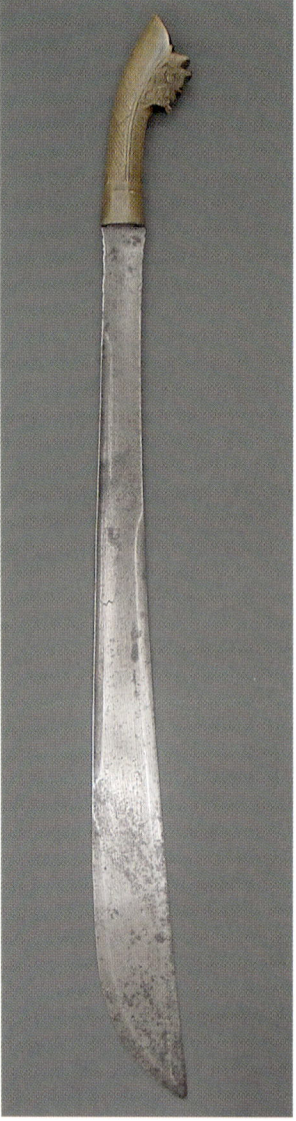

87. Co jang
Sumatra, Aceh. Hilt: *hulu tapa guda*.
L. 77 cm.

88. Co jang
Sumatra, Aceh. Hilt: *hulu tapa guda*.
L. 66 cm.

COAT-OF-MAIL

KALIMANTAN
On Kalimantan a type of coat-of-mail is made of tree bark to which small plates of polished *nassa* and *balush* shells are attached. In north Kalimantan such coats were worn by pirates.
(AVÉ; KEPPEL 1846A)

COLI
[TJOLI, TJOLIKA]

JAVA
A weapon carried by the *apsaras*, divine nymphs.
(HAZEU; KREEMER)

COMBONG
[TJOMBONG]

SUMATRA
A *rencong* with an open-worked blade.
(STONE)

CONDRE
See *chunderik*

CORIK
See *rudus*

CREESE
See *keris*

CRIS
See *keris*

CUDRE
[TJOEDRE]
A type of sword or knife.
(STONE)

CUIRASS
SULAWESI, TORAJA
In combat the Toraja used to wear cuirasses made of tree bark or finely plaited rattan, covered with buffalo hide and laced together on the back. On the front two extensions were attached. Shells polished flat and measuring 5-8 cm serve as decorations.
(ZAAL)

CUNDRIK
See *chunderik*

DAGNE
See *dange*

DAMAOQ
SUMATRA
A small dart for a blow-pipe.
(STONE)

DANDAN
[DENDON]
A *dujung* (sea-cow) tusk used for making hilts. Known as *dandan* (Aceh) or *dendon* (Gayo).
(KREEMER)

DANGE
[DAGNE, DAUGHI]
NIAS
A long, hexagonal wooden shield in the shape of the *kliau* of Kalimantan. The long sides run parallel, the upper and lower sides taper slantingly to a blunt point. This shield has two handles located above each other, across the longitudinal axis. The length is c.150-160 cm, the width c.30-40 cm. It weighs c.4-6 kg. It can be covered with leather or bark and decorated with a stylised human face, geometrical figures or other motifs.
(FELDMAN 1990; HEIN 1890; STONE)

DAUGHI
See *dange*

DENDA
JAVA
A club used as a weapon by gods and heroes in ancient mythology and historical literature.
(RAFFLES 1817; STONE)

DENDON
See *dandan*

DIWAL
JAVA
An arrow used by the mythical gods and heroes.
(RAFFLES 1817)

89. **Dange**
Nias. Wood, covered with leather.
L. 155 cm, W. 38 cm.

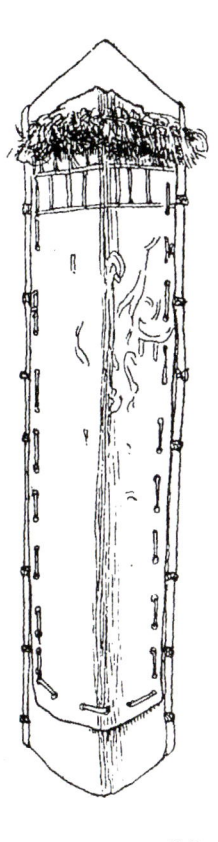

90. **Dange**
North Nias. L. c. 160 cm.

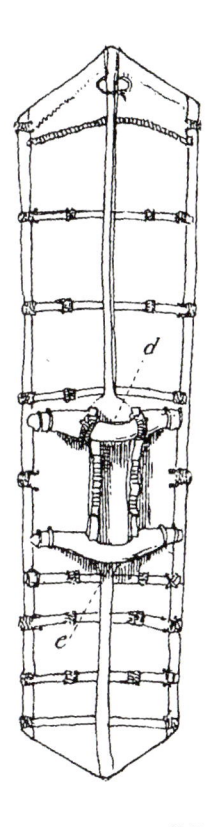

91. **Dange**
Rear view of illus. 90.

DJONAP
See *kalasan*

DJONAP DAIRI
See *jonap dairi*

DJULUNG DJULUNG
See *andar andar*

DOHA
See *dohong*

94. **Dohong**
Kalimantan. L. 44 cm.

95. **Dohong**
Kalimantan. RMV 16-285. Acq. in 1861 from dr. S. Müller. L. 48.5 cm.

DOKE
SULAWESI, TORAJA
Doke is the Toraja term for spear.
(DRAEGER)

DOKE KADANGAN
SULAWESI, TORAJA
A spear with a point the lower part of which broadens on both sides.
(DRAEGER)

DOKE LEPANG
SULAWESI, TORAJA
Spear with a lancet-shaped point.
(DRAEGER)

DOUBLE KNIFE
SUMATRA, BATAK
A short, double knife: two knives carried in a single scabbard.
(SIBETH)

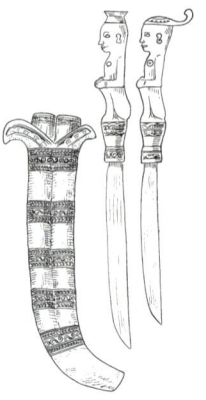

92. **Dohong**
Kalimantan. RMV 952-1. Acq. in 1893 from H.L.H.W. Telling. L. 66 cm.

DOHONG
[DOHA, DUHONG, DUHUNG]

KALIMANTAN
The traditional double-bladed sword or dagger of pre-*mandau* times. It has a ritual significance during funerals (*tiwah*) and is worn by women when the men enter the village with hunted heads. The blade usually has the shape of an arrow-head with a long shaft. It may also have openworked, symmetrical stylised iron figures. The hilt is cylindrical with a protrusion on the end. The *dohong* has a scabbard and may be decorated with human hair. Its belt can have various amulets such as snuff-bottles, teeth of bear and rhinoceros, bits of jaw-bone, (bear) claws etc.
(AVÉ; FABER; FOY; HEIN 1890; SELLATO; STINGL)

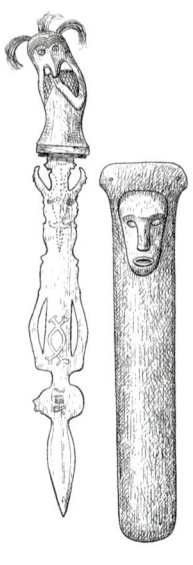

93. **Dohong**
Kalimantan.

96. **Double knife**
Sumatra, Batak.

97. **Double knife**
Sumatra, Karo Batak. RMV 2636-7. Acq. in 1947. L. 28/21 cm.

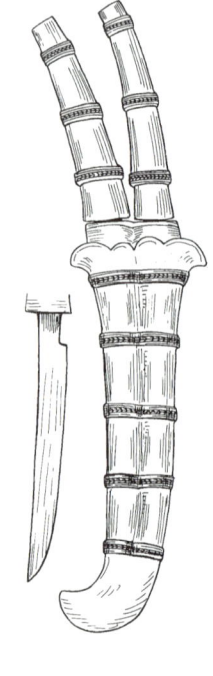

98. **Double knife**
Sumatra, Batak.

DOUBLE SWORD
SUMATRA, BATAK
Two swords kept in a single scabbard. Both swords are of different length.
(FISCHER 1914)

DUA LALAN
SULAWESI, TORAJA
A sword used both in time of war and for the ceremonial slaughtering of buffaloes. Hence its name *dua lalan* (meaning: 'dual purpose'). The blade has an, almost parallel, straight back and edge. At the tip the back curves towards the edge. The short hilt is made of horn, around which brass wire or small strips of rattan are fixed. Its extension is flat and decorated with spiral carving sometimes inlaid with pieces of bone or with mother-of-pearl. The hilt, when looking at its side, may resemble a hornbill's head whereby the beak, the eye and the excrescence on the beak are clearly visible. It can also be more stylised so that only the spiral decorations remain. The straight scabbard has at its lower part a small foot from which it is covered up to 2/3 with rattan, sometimes even completely.
(DRAEGER)

DUHONG
See *dohong*

DUHUNG
See *dohong*

DUKN
KALIMANTAN
A heavy machete in the shape of a German cavalry sabre with a brass cross hilt. The blade is double edged so that the *dukn* can be used to cut and to stab. The scabbard is made of a light variety of wood and painted crimson using dragon's blood (red lacquer). Especially the Undups and the Balaus cover their scabbards and hilts with silver work. A cavity in the hilt is decorated with human hair, the scabbard's rim has a row of feathers from the wings of a hornbill.
(ROTH 1896B; STONE)

DUKU
See *mandau*

DWISULA
JAVA
A two-pointed lance.
(JESSUP)

99. **Dua lalan**
Sulawesi, Toraja. L. 62.5 cm.

100. **Dua lalan**
Sulawesi, Toraja. L. 72.5 cm.

101. **Dua lalan**
Sulawesi, Toraja. L. 64 cm.

102. **Dua lalan**
Sulawesi, Toraja. L. 72.5 cm.

103. **Golok hilt**
Java, Cirebon. RMV 1249-20. Acq. in 1900.

EFG efg

ALPHABETICAL SURVEY

ECCAT
See *piso halasan*

EKAJO
[POKI POKI]

ENGGANO
The *ekajo* is a javelin and the national weapon of Enggano. Male inhabitants of this island exercise javelin throwing from their youth, beginning with short lances measuring less than 1 metre. When sufficiently skilled, they seldom miss their goal within a distance of 50 metres.

The *ekajo* has a large variety of rather blunt points which may be made of iron, copper, bone, bamboo or wood. They usually have a number of barbed hooks on both sides made of one of the above-mentioned materials or, for example, of fish bones or of shark teeth. The number of barbs varies. We often see a different number of barbs on each sides. Their location may well be asymmetric. The wooden shaft becomes thinner towards the end and may be carved with figures. Its top is usually wound around with rope from which sometimes one or more iron spikes pointing downward protrude. No fire is used when manufacturing these metal spikes. The form is reached by patiently beating, grinding and polishing.
(FISCHER 1909; OUDEMANS)

104. **Ekajo**
Enggano. Copper blade.
L. 168 cm, blade L. 12.5 cm.

105. **Ekajo**
Enggano. Iron blade.
L. 204 cm, blade L. 22.5 cm.

EKEH
ENGGANO
A machete made after a Buginese model.
(HELFRICH)

EKKAT
See *piso halasan*

ELTOEP
See *ultup*

ELT'EP
See *ultup*

EMULI
See *salawaku*

ENDONG PANA
JAVA
A quiver tapering somewhat to the bottom, with a broadening, beautifully carved upper half.
(RAFFLES 1817A; STONE)

ENGKAT
See *piso halasan*

ENHERO
MALAKU, BURU
A spear which the Alefuru call *enhero*. In combat its sharp point and shaft (*maen*) are used.
(DRAEGER)

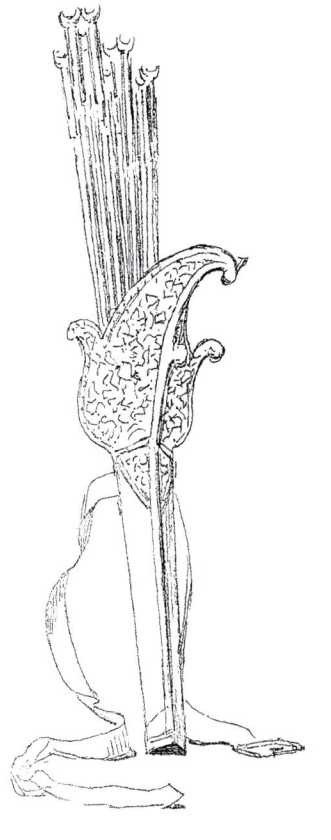

106. **Endong pana**
Java.

FEDJIE
See *feji*

FEDJIE KAIKBARU
See *feji kaikbaru*

FEJI
[FEDJIE]

ENGGANO
A type of machete.
(HELFRICH)

FEJI KAIKBARU
[FEDJIE KAIKBARU]

ENGGANO
A machete made after a Buginese model.
(HELFRICH)

FODA
See *balato*

GABHA
See *akar bahar*

GADA
JAVA, SUNDA ISLES, TIMOR
A club known from Indo-Javanese mythology as a weapon of the ancient gods, demigods and heroes. The *gada* is heavy and has four round balls separated from each other by star-shaped intermediate parts. It may also be a club with a, sometimes oval shaped stone attached to the shaft, sometimes ring shaped and fastened around the shaft.
(DRAEGER; GARDNER 1936; RAFFLES 1817A)

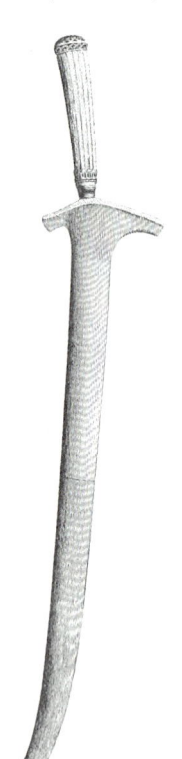

GADA LIMBANG
WEST SUMATRA
A knuckle-duster.
(GARDNER 1936)

GADENG
See *gading*

GADING
[GADENG]
Ivory. The Aceh term is: *gadeng*. See also: *anggang gading*.

107. **Gada** Java.

GADOOBANG
[GADUBONG]
SUMATRA
A sword with a straight back and a slightly S-shaped curved edge broadening at the hilt. At a short distance from the hilt one finds a small protrusion on the sharp edge. The edge curves at the end towards the back into a sharp point. The hilt is slightly curved and broadens towards the end. The *gadoobang* has a scabbard with a strikingly broad segment near the opening.
(DRAEGER; MARSDEN)

108. **Gadoobang** Sumatra.

GADUBONG
See *gadoobang*

GAGONG
[SIMONG]
KALIMANTAN, SEA DAYAK, KENYAH, KAYAN
A war jacket, used in battle and made of animal hide. An incision is made creating a hole through which the warrior places his head in such a way that the animal's head is located on his stomach. Slung over the shoulders and covering his back, it may reach down to his knees. Only the abdominal region and a part of the skin of the forelegs is removed from the hide. Its head is often covered with a piece of metal or a large mother-of-pearl shell (*blasung*) to protect the warrior's stomach. The shell is sometimes located a little lower, below the animal's head. It is said that the shell serves as a boat built by the fallen warrior's spirit to take him across the river to the land of the deceased. The back may be decorated with hornbill feathers, a decoration only meant for a warrior who has won his spurs. Beadwork may be used to decorate it.
This *gagong* is not used often: suitable hides are difficult to find. Skins of goats are preferred due to the long shoulder hair. Black hides are preferred above white ones. Skins of bear, dog, or panther are also used. The jacket is worn more because of its martial appearance than its protective qualities. It may hold back a wooden javelin, but certainly not the thrust of a lance. The Kenyah warrior wears hornbill beaks (*Bucerodtidae*) in pairs on the chest of the *gagong* indicating how many heads he has successfully hunted with his own hands: one pair of beaks for each victim.
(CHIN; FELDMANN 1985; FURNESS; GOMES; HOSE; LOW; RODGERS; ROTH 1896B; STONE)

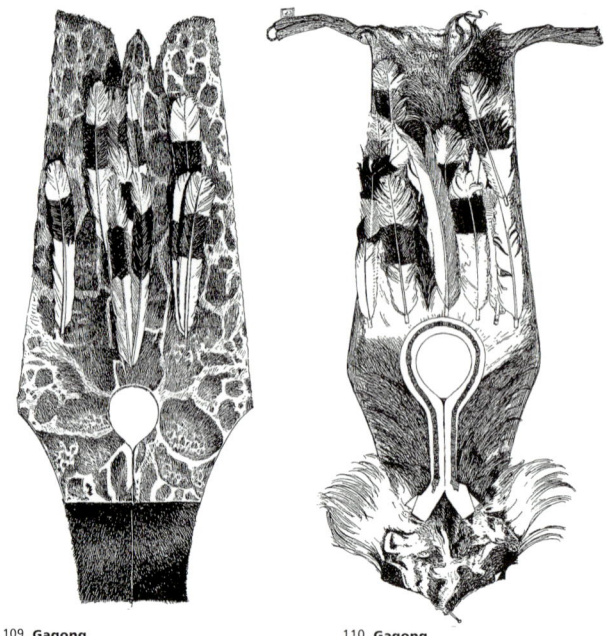

109. **Gagong** Kalimantan. Skin of the *Riaman dahau* (tortoise shell leopard).

110. **Gagong** Kalimantan, Sarebas. Skin of a goat.

GAJANG
SULAWESI, BOLAANG MONGONDO
A sword with a length of *c.*60 cm. The blade is slightly curved. Its edge is located on the inside of the curve. The blade ends in a sharp point. The *gajang* has a simple scabbard.
(FOY)

GALA
MALUKU
A combat staff made of metal or wood.
(DRAEGER)

GALAGANJAR
RIAU
A spear indicating the bearer's rank.
(GARDNER 1936)

GALAIJANG TOKONG
SUMATRA, ACEH
A *sikin panjang* with a rounded point.
(NIEUWENHUIZEN)

GANDJUR
See *tumbak*

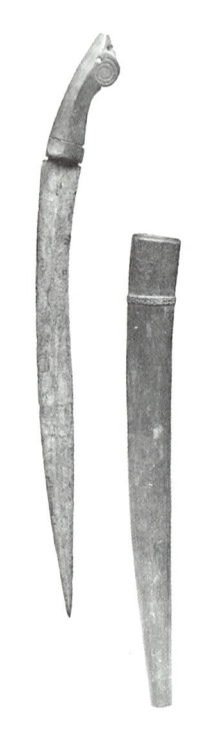

111. **Gajang** North Sulawesi, Bolaang Mongondo.

GANJING
JAVA
An iron bar, used as a weapon prior to 1817.
(RAFFLES 1817A)

GARI
NIAS
A sword with a narrow blade, slightly curved at the end. The hilt has the shape of a *lasara*'s head and a long curved iron protrusion ('tongue'), appearing from the centre of the opened mouth. The scabbard is, as is the blade, slightly curved at the end. It may be decorated with brass strips and wood-carvings. Magical objects may be attached to the scabbard's top. See also *balato*.
(FELDMAN)

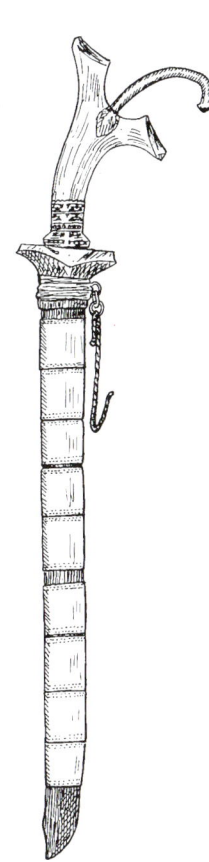

112. **Gari**
Nias. RMV 1239-308. Acq. in 1899 from the widow of dr. H.C.A.E. Helmkampf. L. 57.5 cm.

113. **Gari**
Nias.

GARI TELEGU
See *balato*

GAYONG
A club with magical powers. When struck once by this mythical club, one is wounded in two places. If the blow misses, one is nevertheless injured in one place.
(GARDNER 1936)

GEDUBAN KLEWANG
A short, heavy *klewang*.
(DIELES)

GEGANIT
A spear the point of which is made of a European bayonet. *Geganit* is a corruption of the Dutch word 'bajonet'.
(GARDNER 1936)

GELOEBANG
See *geudubang*

GELUBANG
See *geudubang*

GENDAWA
[GONDEWA]
JAVA
A bow (made of a single piece of wood) often used in combat, but out of fashion since the end of the 17th century, except for state ceremonies. Both ends have long tips made of horn. In the middle of the bow, we see a large wooden grip.
(RAFFLES 1817; SNOUCK HURGRONJE 1892; STONE)

GERPOEH
See *geureupoh*

GERPU
See *geureupoh*

GEUDANG
See *parang geudang*

GEUDOEBANG
See *geudubang*

GEUDUBANG
[GELOEBANG, GELUBANG, GEUDOEBANG, KEDOEBANG, KEDUBANG]
SUMATRA, ACEH
A machete used in the past in combat.
(KREEMER)

GEUREUPOH
[GERPOEH, GERPU]
SUMATRA, ACEH, GAYO-LANDS
A *rudus* with a thick point.
(KREEMER)

GI-GHET
SULAWESI, MINDANAO (PHILIPPINES), SUBUNANS
A bow string.
(STONE)

GLIWANG
See *klewang*

GLIWANG TJOT JANG
See *co jang*

GLOEPAK SIKIN
See *sikin panjang*

GLUPAK SIKIN
See *sikin panjang*

GOBANG
SUNDA ISLES
A type of sword.
(SNOUCK HURGRONJE 1892)

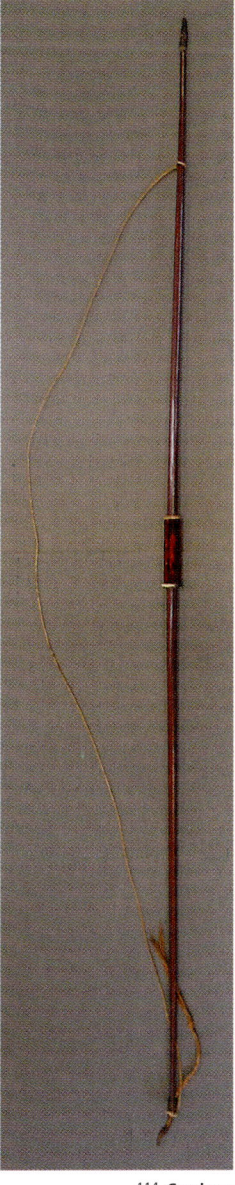

114. **Gendawa**
Java. L. 183 cm.

GOEPUK
See *gupuk*

GOEPUK PERAGIT
See *gupuk peragit*

GOLANG
See *golok*

GOLLOK
See *golok*

GOLOK
[GOLANG, GOLLOK, GOLONG, GOLOQ]

The term *golok* is applied to a variety of machetes once found throughout the Indonesian archipelago. To a certain extent they are still in use today. The similarity between *goloks* is the rather short, firm blade with a straight or somewhat concave back. Especially in the Sunda region of Java, it was the most popular machete and mainly used in combat. The blade is forged in a simple manner. The edge is convex. The hilts and scabbards have various shapes and may be made of numerous materials. In general, they are crudely carved out of wood.

Goloks kept in collections are often the finely made ones. They usually originate from Cikeruh, Sumedang, Sukabumi, Bandung, the Preanger region, Cigasong (Residency Cirebon) and Pahang, all located on Java. Fine *goloks* from Pahang are said to date from the 16th century.

On Java the blade has two main forms:

(a) with a straight or slightly concave back (presumably the original shape). The edge is slightly convex in such a way that the centre of gravity lies between the blade's middle and end in order to deliver heavier blows. The edge curves to the back at the tip. The blade's sides are made smooth;

(b) with a straight or slightly concave back and a slightly or strongly convex edge, too. The back and edge, however, come together in a sharp point. Along the end near the tip, the back is somewhat sharpened. This type of *golok* can thus be used both as machete and a dagger. Both sides of the blade are somewhat concavely shaped to within a few centimetres from the hilt. In some cases we see a blood channel. On these sides figures are carved according to fixed patterns. One side of the blade often shows the name of the village of provenance near the hilt (Cikeruh, written as Tjikeroeh, and often abbreviated Tjike). The year of manufacture may be included. The precise origin and age are thus known, a phenomenon not often occurring with Indonesian weapons.

Blades of type (a) have head-shaped hilts which, in the most plain version, are completely smooth. Finer examples have hilts carved in the shape of a bird's head or *wayang* figure. Blades of type (b) always go with a hilt in the shape of a stylised bird's head.
The hilts are usually made of wood or horn.

Furthermore, a more 'European' type of *golok* is found with a brass parrying piece and a different scabbard. The parrying piece of this type often has a shell-shaped brass decoration on one side. When the weapon is slid into the scabbard, it remains visible on the outside.

The scabbard is usually made of wood but sometimes entirely of horn. Its halves are held together by means of strips of horn, metal or plaited rattan. In addition, it has an upper part, usually made of the same material as the hilt which is slid over its halves and fixed by means of glue, or small pins. The upper part protrudes somewhat on the edge side.

On the scabbard belonging to the bird's head hilt of type (b), the lowest part of the scabbard (sometimes made of horn) is decorated on one side with stylised feathers. This motif can also occur on the scabbard's uppermost part which has an attachment for carrying it on a belt.

In Bandung and Sukabumi, the halves of the scabbard were held together by means of iron-wire. Such scabbards were also made in Europe and exported to Indonesia. In the southern part of the Preanger Residency, scabbards were produced with bone inlays where the sides of the scabbard halves meet. Apart from wooden scabbards, we also find leather ones, often with a brass upper and lower part.

(BEIDATSCH; DIELES; DRAEGER; FORMAN; FOY; GARDNER 1936; HEIN 1899; HELFRICH; HILL 1970; JASPER 1904; JASPER AND PIRNGADIE 1930; MJÖBERG, RAFFLES 1817A; SNOUCK HURGRONJE 1892; STONE; VOSKUIL 1921; VAN ZONNEVELD 1990)

115. **Golok**
Java, Cirebon. RMV 1249-20. Acq. in 1900. L. 57.5 cm.

116. Golok
Java, Cikeruh. L. 31.5 cm.

117. Golok
Java, Cikeruh. L. 49 cm.

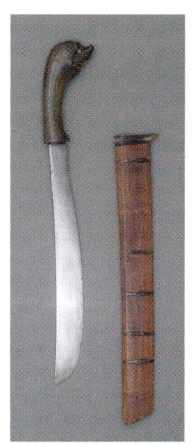

118. Golok
Java. L. 50 cm.

119. Golok
Java. L. 50.5 cm.

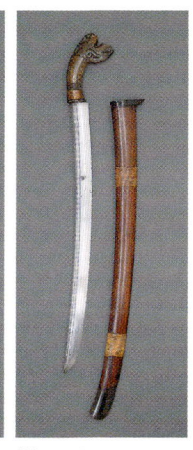

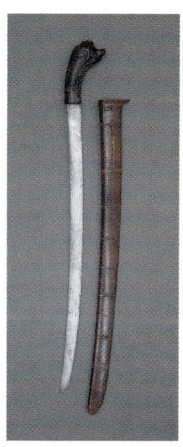

120. Golok
Java. L. 67.5 cm.

121. Golok
L. 66 cm.

122. Golok
Java. L. 30.5 cm.

GOLOK BANGKONG
[GOLOK PERAK]

A type of machete. The blade's broadest part is located in the centre.
(GARDNER 1936)

GOLOK PERAK
See *golok bangkong*

GOLOK REMBAU
SUMATRA, SEMBILAN

A knife in the shape of the *tumbok lada*, in a larger version.
(HILL 1970)

GOLOK TAKA
See *kalasan*

GOLONG
See *golok*

GOLOQ
See *golok*

GONDEWA
See *gendawa*

GONTAR
SUMATRA, MINANGKABAU

A short wooden instrument used to beat the village drum, but also used as a club.
(DRAEGER)

G-OROBANA'A
NIAS

A cuirass made of hard, broad, vertical pieces of buffalo leather widening in the lower part and with an open front. The shoulders are also wide. The seams where the pieces are sewn together form ridges. See also *baru oroba*.
(FELDMAN 1990)

GOROPOH PASE
See *klewang tebal hujong*

GRANGGANG
JAVA, KALIMANTAN, DAYAK

A wooden spear with a straight, rounded point which the Dayak use as a javelin.
(STONE)

GUPUK
[GOEPUK]

SUMATRA, BATAK

A weapon of the Asahan Batak.
(ROGERS)

GUPUK PERAGIT
[GOEPUK PERAGIT]

SUMATRA, BATAK

A weapon of the Asahan Batak.
(ROGERS)

GURU KNIFE
See *lopah petawaran*

123. Hung
Kalimantan. RMV 614-62. Acq. in 1887 from S.W. Tromp. H. 20 cm.

HIJ hij

ALPHABETICAL SURVEY

HALASAN
See *kalasan*

HALBERD
JAVA
On reliefs of *candi* Borobudur (8th-9th century) and *candi* Prambanan (late 9th-10th century) ancient weapons, including halberds, are depicted.
(DRAEGER)

HAMPANG HAMPANG
[AMPANG AMPANG]

SUMATRA, KARO BATAK
An ancient type of shield which, according to Müller, was almost out of use by 1893. It is almost rectangular, tapering a little at the bottom. The long sides may be somewhat concave. The *hampang hampang* is made of buffalo hide and decorated with tassels of *ijuk* fibre and/or horse's hair. These tassels are attached to the top edge and to the upper half of both sides. In the centre a firm handle is placed horizontally. In time of war the rim was also decorated with white feathers.
(FISCHER 1914; MÜLLER 1893; SIBETH)

124. **Hampang hampang**
Sumatra, Batak. L. 65 cm (without decoration).

HANGAN
A weapon from Sumatra.
(ROGERS)

HAROK HAROK
See *sunti abu*

HAUT NYU
See *piso raout*.

HELMET
Numerous types of helmets can be found all over the archipelago. Their forms, materials and decorations are quite diverse.
(a) Amongst the Toba Batak on Sumatra a woven helmet is used. It is a tight fitting, semi-round cap of rattan rings firmly tied together with finely woven thin rattan. These helmets may be decorated with, for instance, beadwork, shells and feathers.
(SIBETH)

125. **Helmet**
Sulawesi. Dutch brass helmet, 17th century, used by the Tobela.

(b) In the Gayo region (Sumatra) a helmet made of buffalo leather is used. The shape is cylindrical, curving to an almost horizontal top. It has a flat extension in the middle at the top.
(VOLZ 1909, 1912)

(c) On Nias helmets or combat hats have a semi-spherical shape. They are made of rattan rings held together by rattan strips. Just below the upper rim a round, firmly woven protruding rim is found. On the sides we see two horizontal protrusions to which a large 'false beard' (made of fibres of the *aren* palm) is attached, giving the entire object a terrifying expression.
(FELDMAN)

(d) On north Sulawesi we find two types of helmets:
(1) of European origin, often dating from the 17th century (*paseki*). In general, these are brass helmets following Spanish models of the VOC (Dutch East Indies Company, 1602-1795);
(2) made of plaited rattan, covered with resin for more strength.
(DRAEGER; GRUBAUER; VOSKUIL)

(e) The Toraja of Sulawesi have helmets which protect and serve to impress the enemy. They are richly decorated and often have stylised buffalo horns of metal as ornamentation.
(GRUBAUER; VOSKUIL)

HEMOLA
TIMOR, SAVU
A sword with an almost rectangular upper part at the hilt. The upper part of the scabbard is also rectangular. The straight blade is rather slender. See also the entry: Swords of the Timor group.
(TEXT: K.H. SIRAG)

126. **Hemola**
Savu. RMV 16-272. Acq. in 1861 from dr. S. Müller. L. 87 cm.

HINA
SUMATRA
A blow-pipe made of two bamboo tubes joined together.
(STONE)

HOEDJOER
See *hujur*

HOJIUR
SUMATRA, BATAK
A spear with a point usually made of iron, sometimes points of bamboo also occur.
(STONE)

HONGKIAM KEK
JAVA, MADURA
A halberd in the form of a spear with a crescent on one side of the point.
(DRAEGER)

HUDJUR HINANDJAR
See *hujur*

HUI THO
SULAWESI, UJUNG PANDANG
A sharp piece of metal spun around on a piece of rope measuring c.1 metre.
(DRAEGER)

HUJUR
[HOEDJOER, HUDJUR HINANDJAR, HUJUR HINANJAR]
SUMATRA, BATAK
A lance with a blade with straight sides tapering towards a blunt point. On both sides the blade has a lengthwise ridge. The *hujur hinanjar* has a shaft decorated by means of rings.
(FISCHER 1914; VAN DER TUUK)

HUJUR HINANJAR
See *hujur*

HULU BABAH BUYA
[HULU SERAMPANG, OELEE BABAH BOEJA, OELOE SERAMPANG, SOEKOEL NGANGO, SUKUL NGANGO]
SUMATRA
A hilt shaped as a 'mouth of a crocodile', a 'harpoon' or a 'mouth wide open'. This hilt is characteristic of the *amanremu* of Gayo and makes a slight angle towards the top ending in two parallel protrusions. Known as: *hulu babah buya* (Aceh), *hulu serampang* (Gayo) and *sukul ngango* (Alas).
(KREEMER)

HULU BOH GLIMA
[HULU GLIMO, OELEE BOH GLIMA, OELOE GLIMO]
SUMATRA
A hilt shaped 'as a pomegranate' (*glima makah*), sometimes used on the *sewar*. The hilt is a lengthened egg-shape, with long notches around the top, and stands at a slight angle to the blade. Known as: *hulu boh glima* (Aceh) and *hulu glimo* (Gayo).
(KREEMER)

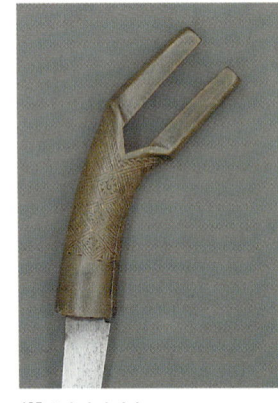

127. **Hulu babah buya**
Sumatra, Aceh. *Amanremu* hilt. L. 15 cm.

128. **Hulu boh glima**
North Sumatra. *Sewar* hilt.

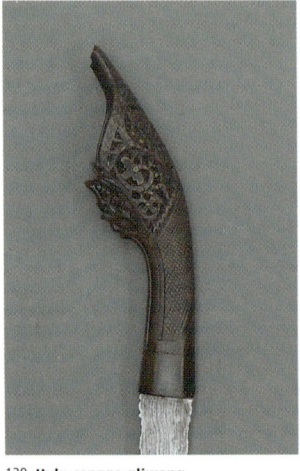

129. **Hulu cangge gliwang**
Sumatra, Aceh. *Rudus* hilt. L. 18.5 cm.

130. **Hulu cangge gliwang**
Sumatra, Aceh. *Rudus* hilt. L. 21 cm.

HULU CANGGE GLIWANG
[HULU COLO, OELEE TJANGGE GLIWANG, OELOE TJOLO]
SUMATRA
A slightly curved hilt tapering to a sharp protrusion at the back. This hilt is found on the *rudus,* a type of *klewang*. Known as: *hulu cange gliwang* (Aceh) and *hulu colo* (Gayo).
(KREEMER)

HULU COLO
See *hulu cangge gliwang*

HULU DANDAN
[HULU DENDON, OELEE DANDAN, OELOE DENDON]
SUMATRA
A hilt 'of white bone' (tusk of a *duyung*, sea-cow) as found on some *rencong*s. Known as *hulu dandan* (Aceh) or *hulu dendon* (Gayo).
(KREEMER)

HULU DENDON
See *hulu dandan*

HULU GEREPUNG
See *hulu puntung*

HULU GEUREUPONG
See *hulu puntung*

HULU GLIMO
See *hulu boh glima*

131. **Hulu dandan**
North Sumatra. *Rencong* hilt.

132. **Hulu dandan**
North Sumatra. *Rencong* hilt. L. 16.5 cm.

HULU IKU ITE
[HULU UKI N ITI, OELEE IKOE ITE, OELOE OEKI N ITI]

SUMATRA

A hilt shaped 'as a duck tail' occurring on kle-wangs. Known as: *hulu iku ite* (Aceh) or *hulu uki n iti* (Gayo).
(KREEMER)

HULU IKU MIE
[HULU SIMPUL, OELEE IKOE MIE, OELOE SIMPOEL, SOEKOEL SIMPOEL, SUKUL SIMPUL]

SUMATRA

A hilt shaped 'as the (curled) tail of a cat' used, for example, for the *luju alas*. The hilt is smooth and straight, sometimes beautifully carved, with at the end a thickened knob or protrusion turning forward towards the edge's sharp side. Known as: *hulu iku mie* (Aceh), *hulu simpul* (Gayo) or *sukul simpul* (Alas).
(KREEMER)

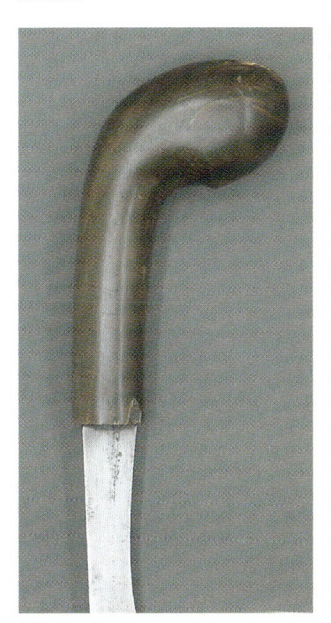

133. **Hulu iku ite**
North Sumatra. *Klewang tebal hujong* hilt.

134. **Hulu iku mie**
Sumatra, Aceh. *Amanremu* hilt. L. 12 cm.

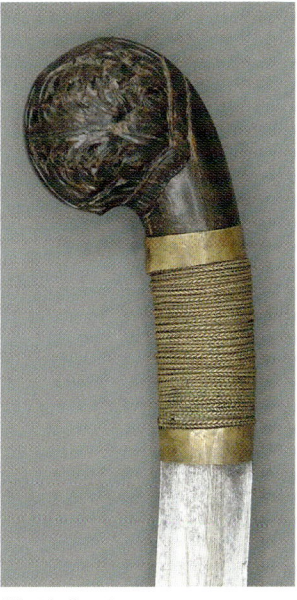

135. **Hulu iku mie**
Sumatra, Batak. *Klewang* hilt. L. 13 cm.

HULU JONGO
[OELEE DJONGO]

SUMATRA, GAYO

A hilt shaped 'as the head of a *jongo*' (a species of stork) found only on the *lopah petawaran* of Gayo. This hilt is thin near the blade, curves in the broader part into a right angle. The hilt ends in a long thin protrusion.
(KREEMER)

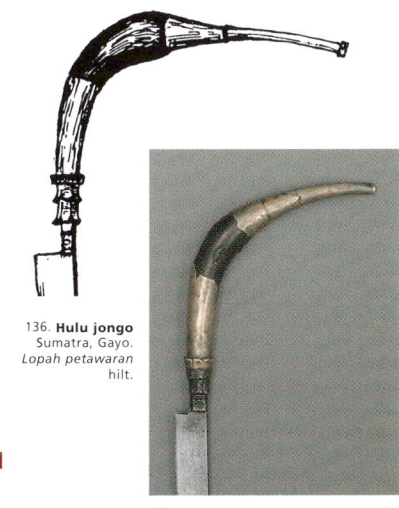

HULU LAPAN SAGI
See *hulu lapan sagu*

136. **Hulu jongo**
Sumatra, Gayo. *Lopah petawaran* hilt.

137. **Hulu jongo**
Sumatra, Gayo. *Lopah petawaran* hilt. L. 12 cm.

HULU LAPAN SAGU
[HULU LAPAN SAGI, OELEE LAPAN SAGOE, OELOE LAPAN SAGI, SOEKOEL LAPAN SAGI, SUKUL LAPAN SAGI]

SUMATRA

An 'octagonal' hilt as with the *sikin lapan sagu*. The hilt is decorated near the blade, the end tapers out into two protrusions in the shape of an open animal mouth. Known as: *hulu lapan sagu* (Aceh), *hulu lapan sagi* (Gayo) or *sukul lapan sagi* (Alas).
(KREEMER)

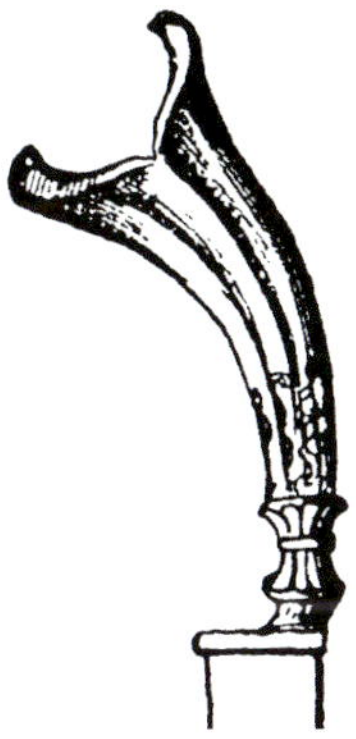

HULU LUNGKEE RUSA
[HULU TANDU N AKANG, OELEE LOENGKEE ROESA, OELOE TANDOE N AKANG]

SUMATRA

A hilt shaped 'as branches from a deer horn'. This is characteristic for the *luju alang* of Gayo. The hilt is slightly curved, with a ring at the base, near the blade. It ends in two short protrusions. Known as: *hulu lungkee rusa* (Aceh) or *hulu tandu n akang* (Gayo).
(KREEMER)

138. **Hulu lapan sagu**
North Sumatra. *Sikin lapan sagu* hilt.

HULU MEU APET
[HULU MUAPIT, OELEE MEU APET, OELOE MOEAPIT, SOEKOEL MEKEPIT, SUKUL MEKEPIT]

SUMATRA

An iron hilt with a basket guard, the hilt of the *peudeueng*, named after the flat small pointed extension on both sides under the basket where the scabbard is fixed. The hilt ends in a bud-like decoration to which a pointed protrusion is attached. The basket's interior is covered with a cushion made of cloth. Known as: *hulu meu apet* (Aceh), *hulu muapit* (Gayo) or *sukul mekepit* (Alas).
(KREEMER)

139. **Hulu meu apet**
Sumatra. *Pedang* hilt. L. 15 cm.

140. **Hulu meu apet**
Sumatra. *Pedang* hilt. L. 14 cm.

141. Hulu meu apet
Sumatra. Hilt of a *pedang*, type I. RMV-3600-410. Acq. in 1959 through the Ethnografisch Museum van de Koninklijke Militaire Academie, Breda.

HULU MEUCANGGE
[OELEE MEUTJANGGEE]

SUMATRA
A hilt with a rectangular 'curve', most often found on *rencong*s. The hilt begins thinly near the blade, broadens to then curve in a right angle to a long and thin straight protrusion broadening towards the flattened end.
(KREEMER)

142. Hulu meucangge
North Sumatra. *Rencong* hilt, decorated with *tampos* at the base. L. 14.5 cm.

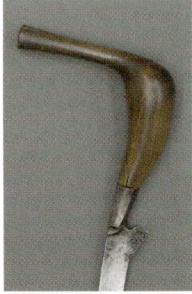

143. Hulu meucangge
North Sumatra. *Rencong* hilt. L. 11 cm.

HULU MUAPIT
See *hulu meu apet*

HULU PAROH BLESEKAN
[OELEE PAROH BLESEKAN]

SUMATRA, GAYO
A hilt resembling the beak of the *blesekan* (a species of snipe) only found on the *luju celiko* in Gayo. The hilt begins narrowly near the blade, broadens towards the top, to then bend sharply and end in a sharp point.
(KREEMER)

144. Hulu paroh blesekan
Sumatra, Gayo. *Luju celiko* hilt.

HULU PEUDADA
[OELEE PEUDADA]

SUMATRA
A type of hilt occurring on the *sikin panjang*.
(ROGERS)

HULU PEUSANGAN
[OELEE PEUSANGAN]

SUMATRA, ACEH, PEUSANGAN
A hilt which broadens at the top, as used with the *sikin panjang* or with the *pedang*. The hilt is almost straight and ends in two flattened protrusions, at the tip broad enough to almost touch each other. This form is characteristic of Peusangan in north Aceh.
(KREEMER; VOLZ 1912)

145. Hulu peusangan
Sumatra, Aceh. *Sikin pasangan* hilt. L. 16 cm.

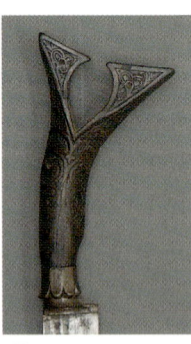

146. Hulu peusangan
North Sumatra. *Sikin pasangan* hilt. L. 15 cm.

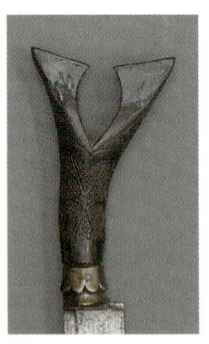

147. Hulu peusangan
North Sumatra, Peusangan. *Sikin panjang* hilt. L. 14 cm.

HULU PUNTUNG
[HULU GEREPUNG, HULU GEUREUPONG, OELEE GEUREUPONG, OELOE GEREPONG, OELEE POENTONG, OELOE POENTONG, SOEKEL GERPONG, SUKUL GERPONG]

SUMATRA
A hilt with a truncated end, sometimes plain, sometimes forked, with carved leaf tendrils at the top as may be seen on *rencong*s. Known as: *hulu puntung* or *hulu geureupong* (Aceh), *hulu puntung* or *hulu gerepung* (Gayo) and *sukul gerpong* (Alas).
(KREEMER)

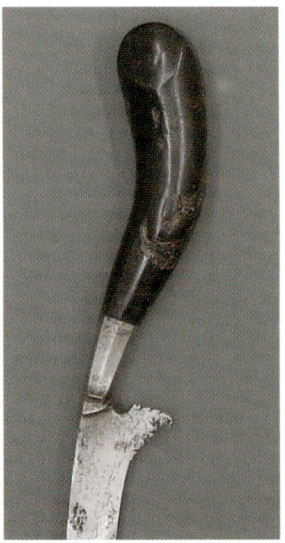

148. Hulu puntung
North Sumatra. *Rencong* hilt. L. 11 cm.

149. **Hulu puntung**
North Sumatra. *Rencong* hilt. L. 14 cm.

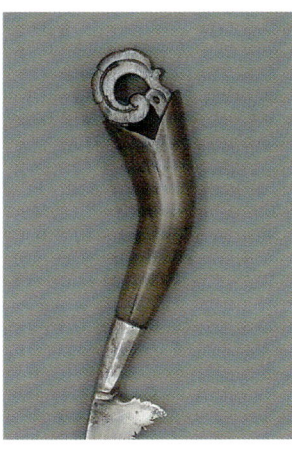

150. **Hulu puntung**
North Sumatra. *Rencong* hilt. L. 14 cm.

HULU RUMPUNG
[OELEE ROEMPOENG]

SUMATRA
A hilt at the end of which a shallow V-shape is found. This type of hilt often occurs on the *sikin panjang* and the *luju alang*.
(KREEMER)

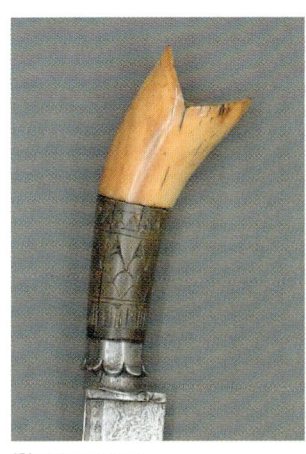

151. **Hulu rumpung**
North Sumatra. *Sikin panjang* hilt. L. 13 cm.

152. **Hulu rumpung**
North Sumatra. *Luju alang* hilt, intensively used. L. 11 cm.

HULU SERAMPANG
See *hulu babah buya*

HULU SIMPUL
See *hulu iku mie*

HULU TANDU N AKANG
See *hulu lungkee rusa*

HULU TAPA GUDA
[HULU TAPA KUDO, OELEE TAPA GOEDA, OELOE TAPA KOEDO, SOEKOEL TAPA KOEDO, SUKUL TAPA KUDO]

SUMATRA
A hilt resembling the foot of a horse is characteristic of the *ladieng*, the *rudus* or the *co jang*. The hilts of the *ladieng* are usually less ornate and often turn slightly backwards at the top. Known as: *hulu tapa guda* (Alas), *hula tapa kudo* (Gayo) or *sukul tapa kudo* (Alas).
(KREEMER)

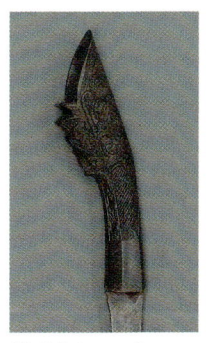

153. **Hulu tapa guda**
Sumatra, Aceh. L. 18 cm.

154. **Hulu tapa guda**
Sumatra. *Klewang* hilt. L. 15 cm.

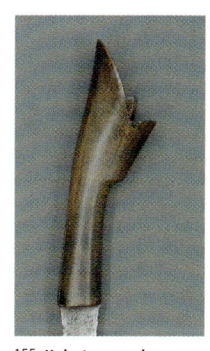

155. **Hulu tapa guda**
Sumatra, Aceh. *Ladieng* hilt. L. 19 cm.

156. **Hulu tapa guda**
Sumatra. *Klewang* hilt. L. 15 cm.

157. **Hulu tapa guda**
Sumatra, Aceh. *Co jang* hilt. L. 15 cm.

HULU TAPA KUDO
See *hulu tapa guda*

HULU TUMPANG BEUNTEUENG
[OELEE TOEMPANG BEUNTEUENG]

SUMATRA
A hilt with a deeply cut out V-shaped end. Hilts such as the *hulu tumpang beunteueng* are found on the *sikin panjang* and the *lading*.
(KREEMER)

HULU UKI N ITI
See *hulu iku ite*

HUMARANGA
HALMAHERA, TOBELO

158. **Hulu tumpang beunteueng**
North Sumatra. *Sikin panjang* hilt. L. 15.5 cm.

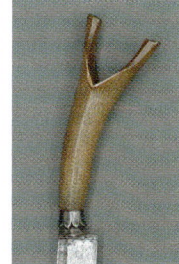

159. **Hulu tumpang beunteueng**
North Sumatra. *Sikin panjang* hilt. L. 18 cm.

A sword with a straight edge and back. The blade broadens towards the tip where the back curves towards the edge. The wooden hilts, usually blackened with soot and plant juice, are carved by the Tobelo themselves. Types in the shape of a bird's beak or a crocodile's mouth (*o gohamanga ma uru*) are found, presumably representing animal ancestors.
The blades of these swords used to be part of spoils or purchased from non-Tobelo traders. Even recently (1990) the blades were not forged by the Tobelo themselves, but by smiths from elsewhere who had settled down in the region in the course of time. The shape is, nevertheless, drawn by the Tobelo, using existing swords as examples.
The term *humaranga* is derived from Semarang, a city on Java, from which this type of swords probably originates. They were confiscated from enemies to be imitated again and again. Their

foreign provenance is expressed in the hilts with motifs made of material from overseas, such as pieces of mother-of-pearl shell from the *nautilus* shell or pieces of broken earthenware. The *humaranga* may be part of the ceremonial exchange of gifts at a wedding. It is worn during the *hoyla* dance, the war dance performed during the marriage ritual.
(PLATENKAMP)

HUNG
KALIMANTAN, DAYAK
A small gourd (*Lagenaria vulgaris*) carried by the Dayak to store small amounts of tree pith which forms the back end of blow-pipe darts. The gourd is closed by means of a wooden lid often decorated with a carved monster's head.
(STONE; WEIGLEIN)

160. **Hung** Kalimantan.

161. **Hung** Kalimantan. RMV 614-62. Acq. in 1887 from S.W. Tromp. H. 20 cm.

HWA KEK
JAVA, MADURA
A halberd in the shape of a spear. Its point has a crescent on both sides.
(DRAEGER)

IKAN PARI
A sting-ray. In early times fishermen used the poisonous tip of a sting-ray's tail as a weapon (spear-point or dagger), or to manufacture poison. *Ikan pari* tails have been found in caves, sometimes located at a distance of five to six days' journey from the coast. Apparently these poisonous, notched sting-ray tails proved to be a good weapon. Its hilt must have been covered with tree bark or some other material to protect the hand. Once the poison had lost its power, poison of a freshly caught *ikan pari* was applied. The short, straight form of the oldest known *keris* type, the *keris majapahit*, was perhaps derived from an *ikan pari* tail.
(GARDNER 1936; SOPHER; VAN DER STRAATEN 1973)

IKOE LINONG
See *parang iku linong*

IKU LINONG
See *parang iku linong*

ILANUN KAMPILAN
A long variant of the *golok* and *gedubang*.
(DIELES)

INDAN
JAVA
A club of the Indo-Javanese deities, demigods and heroes from antiquity. This club is straight with a smooth handle which broadens slightly towards the tip. Its spiral shaped body ends in a blunt point.
(RAFFLES 1817; STONE)

162. **Indan** Java.

INDOEPO
See *indupo*

INDUPO
[INDOEPO]
NORTH SULAWESI
A blow-pipe with a mouth-piece of horn, used only amongst the inhabitants of the coast and mountain regions in the north-west corner of Tomini Bay (Moutong) from Pagur to Parigi.
(LOEBÈR 1928)

IPOH
[SIREN, UPAS]
A poison applied to blow-darts, shot by means of the *sumpitan*, mainly on Kalimantan. The most important ingredients are the juice of the *ipoh* or *upas* tree (*Antiaris toxicara*, Praceae family) and the juice of the roots of the *ipoh*-creeper (*Strychnos tieute*, also known as *ipoh akar* or *ipoh gunong*). The milk-like juice of the *ipoh* tree is harvested by cutting a fish-bone pattern into the bark and then collected in a bamboo tube. The higher up the tree it is found, the more toxic it is said to be. Both ingredients are used separately or in combination. Sometimes they are mixed with substances such as tobacco, (red) pepper, onion, stings of scorpions, venom-fangs of snakes and arsenic. The ingredients and quantities may vary, but *ipoh* juice always forms the basis. The other additions serve an emotional or magical purpose, rather than contributing to its strength. The juice is slowly reduced over a fire to a purple-brown, blackish substance. It should not be overheated, or it will lose its effect. Before being applied to the tip of a dart, the paste is softened on a small wooden palette with a wooden spatula. Once the poison is applied to the darts, these are dried in the sun or over a fire. *Ipoh* instantly kills birds and small mammals when it enters the bloodstream. Larger animals may live for several minutes, the largest apes for a quarter of an hour.
The *Suma Oriental* written by the Portugese traveller Tome Pires in 1512-1515 mentions the use of poisonous darts. So does his fellow countryman Fernao Mendes Pinto in 1555.
(GOMES; HOSE; SKEAT; SOPHER; STONE; WEIGLEIN)

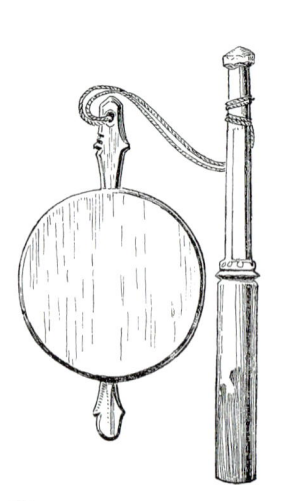

163. **Ipoh plate** Kalimantan, Punan. Circular plate with rolling pin, used for preparing *ipoh* (poison).

ISAU
KALIMANTAN, BAJAU
Legend has it that this fine weapon was made of unusual bits of steel and iron, collected at unusual moments. Once the metal pieces were twisted together, they were forged before being moulded into their final shape. The hilt, made of hardwood or horn, was strengthened and decorated by means of several rings demanded from the inhabitants of a 'longhouse'. Each family supplied at least one copper or silver ring. It is also said that the blacksmith could not ask for a reward for manufacturing an *isau*.
(ROTH 1896A)

JABANG
See *piso raout*

JAMBIA
See *jambiah*

ALPHABETICAL SURVEY

164. Jimpul
Kalimantan. L. 60.5 cm.

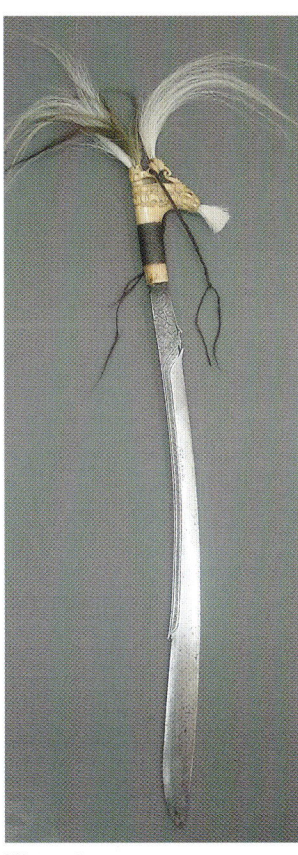

165. Jimpul
Kalimantan. L. 77 cm.

166. Jimpul
Kalimantan. L. 62 cm.

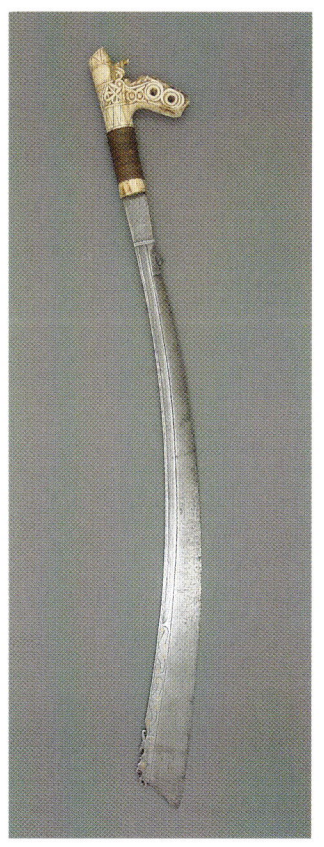

167. Jimpul
Kalimantan. L. 71.5 cm.

JAMBIAH
[JAMBIA, JAMBIYA]

JAVA, SUMATRA, MADURA
A curved knife with a double-edged blade. The broadest part at the hilt tapers with a distinct curve to its point. The blade (often with an elevated ridge along the centre) may well be of Arabian or Indian origin and often has an indigenous hilt. In general, locally forged blades are smaller and without a ridge.
(GARDNER 1936; DRAEGER; STONE)

JAMBIYA
See *jambiah*

JEKINPANDJANG
See *sikin panjang*

JIMPUL
[JUMBUL, PARANG DJIMPUL]

KALIMANTAN, SEA DAYAK, KENYAH
The *jimpul* is an intermediary form between the *parang ilang* and the *langgai tinggang* dating from c.1870. The blade of this machete has flat sides and is distinctly curved. Widening towards the point, it ends in a slanting angle or rounded tip. The edge is longer than the back. The blade may have two or three grooves, running at a short distance from the back, as well as hooks and protrusions (*krowit*) near the hilt on the sharp edge. Chased figures can be found on both sides near the hilt. The hilt and scabbard are made in the same way as those of the *mandau*. Its scabbard is, of course, curved.
(FORMAN; GARDNER 1936; HEIN 1899; SHELFORD; STINGL)

JONAP
See *kalasan*

JONAP DAIRI [DJONAP DAIRI]

SUMATRA, BATAK
A variant of the *kalasan*. Its hilt has the shape of an hour-glass. The scabbard has a rectangular small protrusion at the edge's side near the scabbard's mouth.
(VAN DER TUUK)

JONO

SUMATRA, BATAK
A sword with a slightly curved blade and a straight hilt with a slightly rounded top. The scabbard has a cross-piece with a curved upper part protruding further on the sharp side than on the rear side.
(DRAEGER; MARSDEN)

JOWING

KALIMANTAN, DAYAK
Loose tip of a blow-dart.
(STONE)

JULUNG JULUNG
See *andar andar*

JUMBUL
See *jimpul*

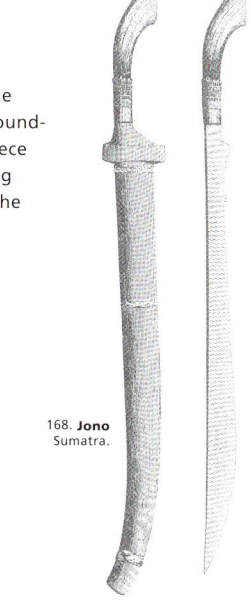

168. Jono
Sumatra.

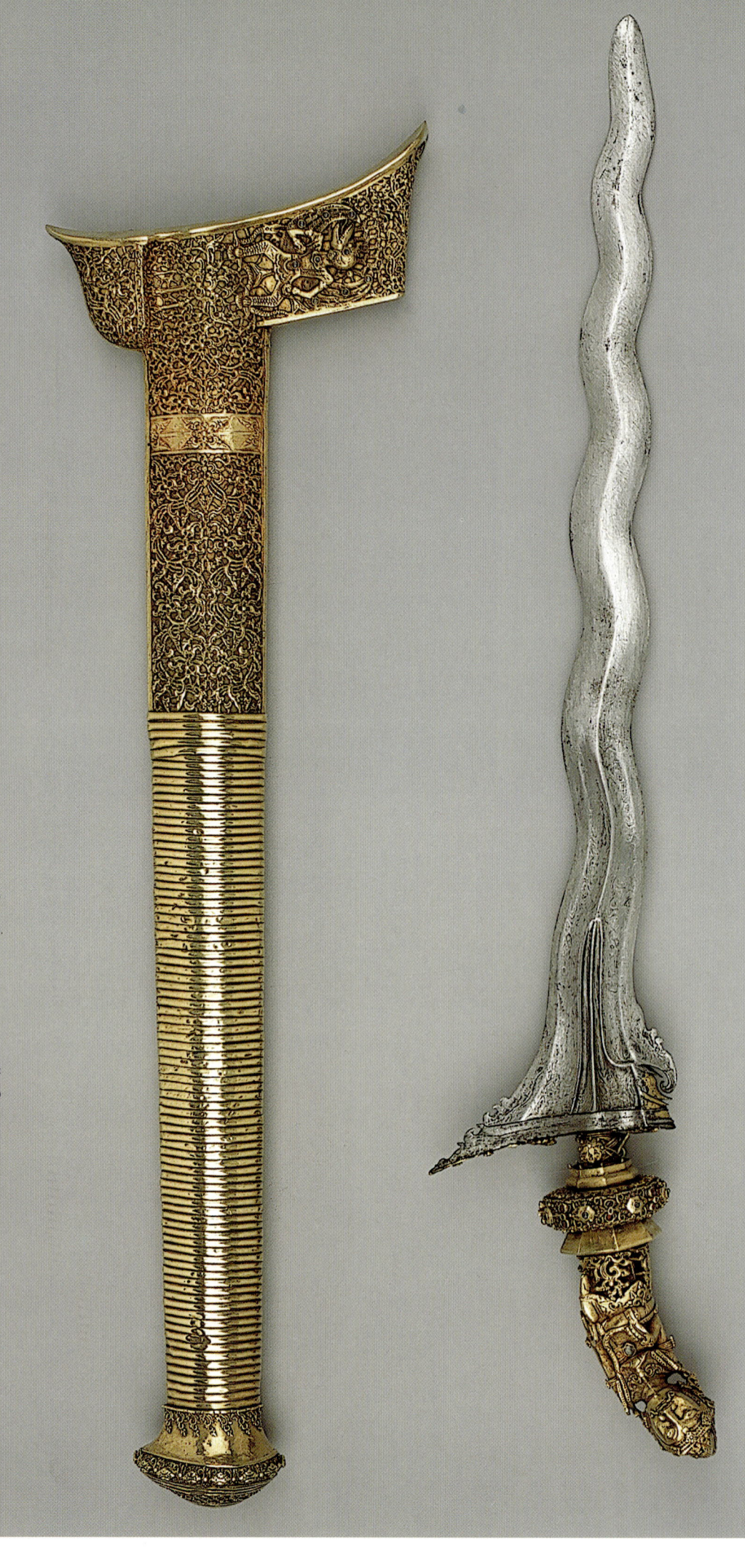

169. Keris
Java/Sulawesi. RMV 360-6021 Acq. in 1883 through the Koninklijk Kabinet van Zeldzaamheden, 's-Gravenhage. Collected before 1750. L. 46 cm.
This keris is unusual not only as an example of perfect workmanship and because elements of the decoration represent the period before the conversion of Sulawesi to Islam but also because of its great documentary importance.
As early as the middle of the eighteenth century this weapon was in the possession of Stadholder Willem IV (1711-1751) in his function as head of the Dutch East India Company (VOC) which he assumed in 1749. It seems probable that this keris had not been made recently and had come into the possession of the Dutch East India Company in connection with the upheavels in Makassar in 1666-1669.
Although this keris originated from the Buginese culture area of south Sulawesi, it is not entirely certain whether it was made by Buginese craftsmen. The gold hilt of the keris which is set in small gem stones, is in the form of a bent wayang figure who, on account of his long thumb nail, is remenicent of the wayang hero Bima. The standing Garuda on the upper part of the scabbard shows a pronounced east Javanese character.
It is no longer possible to tell whether the goldsmith who produced the ornamentation on this weapon did so in east Java after a local model or to suit the taste of the Buginese for whom it was intended, or was himself a Javanese in Sulawesi.
It was the oldest artifact from Indonesia in the possessions of the House of Orange-Nassau which has been preserved.

(VAN DONGEN; VIANELLO; WASSING-VISSER)

K k
ALPHABETICAL SURVEY

KABEALA
EAST SUMBA
A straight-backed sword with a somewhat convex edge. The blade broadens slightly towards the tip. Its back curves towards the edge at the tip. The hilt is solid, curving halfway at an angle of 45°. The scabbard is straight and has a large number of woven strips to keep both parts together. Its mouth has a slanting protrusion towards the blade's sharp side.
(FISCHER)

170. Kabeala
Sumba, Wayawa. L. 50.5 cm.

171. Kabeala
Sumba, Loli district. L. 53 cm.

172. Kabeala
Sumba, Lamboya district. L. 58.5 cm.

173. Kabeala
Sumba. L. 48 cm.

KABHA
See *akar bahar*

KAHUK
TIMOR
A blow-pipe made of bamboo.
(STONE)

KAHUK ISIN
TIMOR
An arrow for a blow-pipe (*kahuk*) made of bamboo. The end is covered with small feathers. These arrows are c.90 cm long, much longer than blow-pipe arrows found elsewhere.

KALAPU
See *katapu*

KALASAN
[ALASAN, DJONAP, GOLOK TAKA, HALASAN, JONAP, KALASEN, KALASSAN, PISO GULAK TAKA, PISO GULUK TAKA, PISO KALESEN, PISO REMPU PIRAK, PISO ROEMBOE PIRAK, PISO RUMBU PIRAK, REMPU PIRAK, ROEMBOE PIRAK, RUMBU PIRAK]

SUMATRA, TOBA BATAK, EAST KARO, GAYO
A sword with a straight or somewhat concave back, a slightly S-shaped or completely straight edge, and a sharp point whereby the edge runs towards the back. Near the hilt the edge has one or more protruding decorations. The hilts vary in shape and may be made of buffalo horn, deer horn, ivory or bronze. The hilts of deer horn and of ivory are short and thick. They either widen broadly towards both points and are thus thinner in the centre, or only widen broadly near the end. The hilts made of buffalo horn have models such as the *sukul nganga*, the *sukul jering*, or the mouth-shaped hilt. The scabbard may be straight, with the mouth widening towards the edge's side, or slightly curved with the end forming a pronounced curve towards the blade's back. The halves of the scabbard are held together by means of plaited rattan or thin or broad metal strips. The term *piso rempu pirak* presumably means 'knife with silver strips'.
(DRAEGER; FISCHER 1914; FORMAN; MÜLLER; PARAVICINI; ROGERS; VAN DER TUUK; VOLZ 1909, 1912)

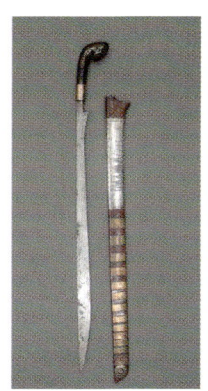

174. Kalasan
Sumatra. Hilt: *sukul jering*. L. 62 cm.

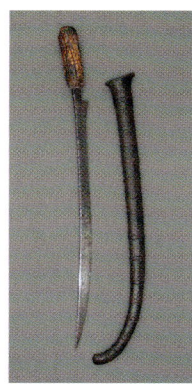

175. Kalasan
Sumatra. L. 56 cm.

176. Kalasan
Sumatra. L. 59 cm.

KALASAN SITUKAS
SUMATRA, EAST KARO
A *kalasan* with a straight back and an S-shaped edge. Its blade is broader than that of the regular *kalasan*. The edge is decorated with various protrusions near the hilt. The scabbard is straight with a mouth which broadens towards the blade's edge.
(FISCHER 1914; FORMAN; VOLZ 1909)

KALASEN
See *kalasan*

KALASSAN
See *kalasan*

KALAVIT
KALIMANTAN, SEA DAYAK
A type of shield.
(ROTH 1896B)

KALEWANG
See *klewang*

KALIHAN
KALIMANTAN, KAYAN
A shield of the normal Dayak type. See *kliau*.
(STONE)

KALIJAWO
[KALIWO, KALIYAWO]

CENTRAL AND SOUTH SULAWESI, TORAJA
A shield the shape of which resembles the *kliau* of the Dayak in Kalimantan.
(HEIN 1890; HEIN 1899; LEENDERTZ)

KALIJAWO MALAMPE
[KALYAWO MALAMPE, LENGOE LABOE, LENGU LABU]

SULAWESI
A long, rectangular shield with rattan strips horizontally across the front. Known as *kalijawo malampe* (Buginese) and *lengu labu* (Makassar).
(MATTHES 1874, 1885; SCHRÖDER)

KALIJAWO MALEBOE
See *kalijawo malebu*

KALIJAWO MALEBU
[KALIJAWO MALEBOE, KALYAWO MALEBOE, KALYAWO MALEBU, LENGOE BODONG, LENGU BODONG]

A round shield known as: *kalijawo malebu* (Buginese) and *lengu bodong* (Makassar).
(MATTHES 1874, 1885; SCHRÖDER)

KALIWO
See *kalijawo*

KALIYAWO
See *kalijawo*

KALUPU

KALIMANTAN, DAYAK
A helmet made of basketry covered with porcupine skin. In the centre the spines slant backwards and come together at the top.
(ROTH 1896B)

KALUS

SUNDA ISLES
A whip measuring c.90 cm with a handle and a flexible part.
(DRAEGER)

KALYAWU MALAMPE
See *kaliyawo malampe*

KALYAWO MALEBOE
See *kalijawo malebu*

KALYAWO MALEBU
See *kalijawo malebu*

KAMPAR SABIT

SUMATRA, MINANGKABAU
A type of sickle.
(DRAEGER)

KAMPILAN
[BADA, BADANUMOGANDI, BARA, CAMPILAN]

KALIMANTAN, SULAWESI, TALAUD ISLES, PHILIPPINES
A long sword usually held in both hands. Generally speaking, the blade has a straight back and edge, diverging from each other towards the point. Here the back slants towards the edge. On the slanting part we often see a protrusion or spike. The hilt has a large upper part. It is often decorated by means of hair (red or black), inlay and carvings. A large symmetrical cross-piece is found at the hilt's base. The scabbard is usually made of two flat pieces of wood or bamboo. Both parts are held together by means of rattan strips, only on the lower part. When the blade is drawn from the scabbard, the two parts bend apart to allow the *kampilan* to be removed. Therefore, the entire scabbard need not be of the same width as the broadest part of the blade.
In the Philippines we find a type of scabbard made of a single piece of wood with a grip in the centre. This scabbard can thus also serve as a shield.
The *kampilan* was originally the national weapon of the Sea Dayak. It was later adopted by the Moros of Sulu and Mindanao (Philippines).
[FORMAN; FOY; HEIN 1899; SHELFORD; STONE]

KAMPING
See *mandau*

KANAKINIE

ENGGANO
A type of lance.
(HELFRICH)

KANJDJAI
See *kanjai*

KANJAI
[KANJDJAI]

SULAWESI
A spear with an asymmetrical point, with a barb on one side. *Kanjai* is both the Buginese and the Makkassar term.
(MATTHES 1874, 1885; SCHRÖDER)

KANTA

CENTRAL SULAWESI, POSO REGION, TORAJA
A long slender shield, V-shaped over its entire length. It tapers somewhat towards the lower and upper parts. It is richly decorated with goat's hair dyed white, black and red, and is inlaid with small shells or white bone.
(GRUBAUER; VOSKUIL 1921)

KAPA
See *leumbeng*

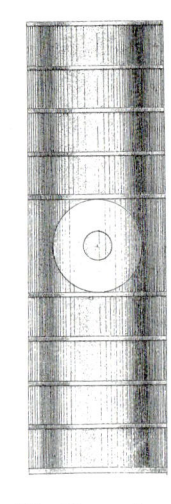

177. **Kalijawo malampe** Sulawesi.

178. **Kalijawo malebu** Sulawesi.

179. **Kampilan** Timor. RMV 351-65. Acq. in 1883 through the Ministerie van Marine. L. 103.5 cm.

180. **Kampilan** Kalimantan (?). L. 98 cm.

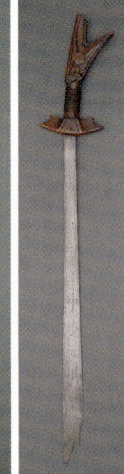

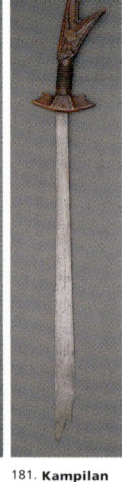

181. **Kampilan** Kalimantan (?). L. 102 cm.

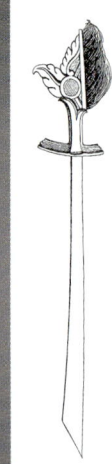

182. **Kampilan** Kalimantan, Lanun.

183. **Kanta** Central Sulawesi. Front and rear view. L. 115 cm.

184. **Kanta** Sulawesi. RMV 3600-5831. Acq. in 1959 through the Ethnografisch Museum van de Koninklijke Militaire Academie, Breda. L. 111.5 cm, W. 16 cm.

KAPAK I
Sumatra, Batak
A (throwing) axe made of iron.
(Draeger; Gardner 1936; Marsden)

KAPAK II
[KAPAQ]

SUMATRA, ACEH

A lance or javelin with a blade which broadens from the point onwards, then becoming narrower and then broadening again to form a round metal point attached to the shaft.
(NIEUWENHUIZEN; SNOUCK HURGRONJE 1904)

KAPAK JEPUN
A combat axe in the shape of a long stick with a hammer-shaped metal head and sharp points on the upper part. Its name presumably refers to its Japanese origin.
(DRAEGER)

KAPAQ
See *kapak II*

KARAMBIT
See *lawi ayam*

KARBAR
See *akar bahar*

KARIS
KALIMANTAN, DAYAK

The Dayak term for *keris* or a similar weapon worn only as an ornament. The blade is straight (*sapukal*) or waved (*parung*). The hilt is made of fine wood, deer horn or bone. It usually finely carved in the shape of a snake's head, a bird's head or some fabled animal.
(HEIN 1890; HEIN 1899; VAN DER STRAATEN 1973)

KAROENKOENG
See *karungkung*

KARUNKUNG
[KAROENKOENG]

KALIMANTAN

A cuirass made of rattan.
(STONE)

KASO
[KASOK, KASOQ]

SUMATRA, ACEH

A long, straight double-edged sword. The hilt is straight and cylindrical in cross-section, broadening towards the point and the blade. In the past those seeking revenge used this weapon when standing under the pile-house and plunging it through the floor on which the master of the house had reclined.
(KREEMER; NIEUWENHUIZEN)

KASOK
See *kaso*

KASOQ
See *kaso*

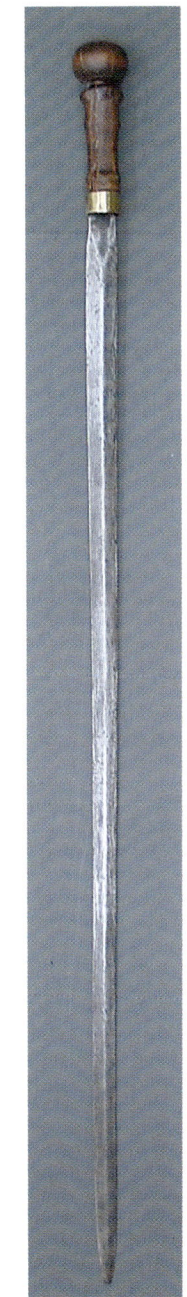

185. **Kaso**
Sumatra, Aceh. L. 104 cm.

KATAPU
[KALAPU]

KALIMANTAN, DAYAK

When the Dayak wage war they wear a semi-circular, tightly fitting woven helmet made of thick rattan, split in two, lengthwise. The *katapu* may have an exactly fitting inner helmet forming an excellent protection against the blow of a sword. This helmet is often covered with metal plates, large scales of fish, skin of bear, of monkey or of other animals. Beadwork serves as a decoration. So do the tail-feathers of poultry, the claws, beak or the head of a hornbill, human hair or other dyed hair, shells, teeth of bear or panther etc. The decoration is sometimes in the shape of a monster's head. The rim can be edged with red flannel and decorated with *nassa* shells. Decorations with hornbill tail-feathers were reserved for warriors who had participated in a successful raid. Each feather may represent a killed enemy. These helmets not only serve to protect, but also to enhance the martial appearance. See also *sampulau anggang*.
(AVÉ; FELDMAN 1985; FURNESS; GOMES; HOOP; HOSE; LOW; ROTH 1896A, 1896B; STONE; VOSKUIL)

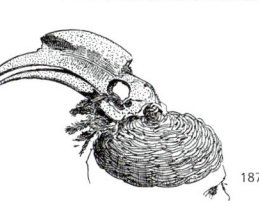

186. **Katapu**
Kalimantan. RMV 401-40. Acq. in 1883 from mr. J.W. Van Lansberge. H. 16.5 cm.

187. **Katapu**
South east Kalimantan. Wickerwork, skull and bill of the hornbill.

188. **Katapu**
South east Kalimantan. Furnished with a.o. the bill, feathers and claws of the hornbill.

189. **Katapu**
Kalimantan. RMV 1573-13. Acq. in 1906 from E. Van Walchren. H. 14 cm.

190. **Katapu**
Kalimantan, Baram River, Dayak. Skin of the porcupine. Diam. 18 cm.

191. **Katapu**
Kalimantan, Lutong Kayan. Plaited rattan with armadillo scales.

KATAPU KALOI
[KATUPU KALOI]

KALIMANTAN, SEA DAYAK
A helmet made of soft, flattened, bark-like material to which large shells are sewn using finely split rattan.
(LING; ROTH 1896B; STONE)

KATOK I
A short skirt made of coloured silk or cotton. This battle dress reaches just below the knee.
(GARDNER 1936)

KATOK II
[PANDAK]
A type of knife
(STONE)

KATUEN
[TODOPENAN]

MALUKU, BURU
A scabbard of the *todo*. In the southern part of Buru, this scabbard is used instead of a shield. The name is derived from *kau* (tree) and *tuen* (stump of a tree).
(DRAEGER)

KATUNGUNG

SUMATRA, BATAK, PAKPAK
A sword with an almost straight edge and back. The point is broader than the rest of the blade. The back curves towards the edge. The hilt, *sukul katungangan*, is thick and bends at 90° at the top, ending in the shape of a mouth with two open lips. The *katungung* is carried in an almost straight scabbard broadening only at the mouth. Its halves are held together with strips.
(VOLZ)

KATUPU KALOI
See *katapu kaloi*

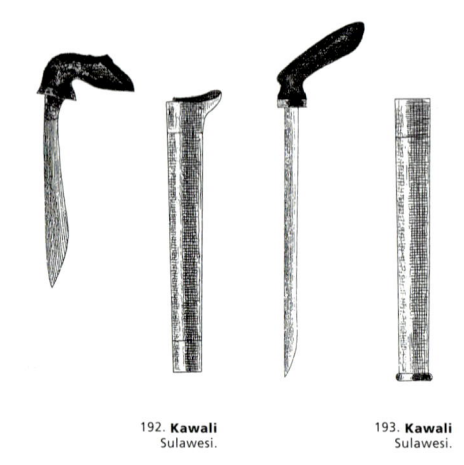

KAWALI
[BADI GOEROE, BADI GURU]

SOUTH SULAWESI
A knife with a right angled hilt. Its slanting tip sometimes has a cross-wise indentation. *Badi guru* means: 'priest's dagger' (Makassar) and is also known as: *kawali* (Buginese). See also *badek*.
(HEIN 1899; MATTHES 1874, 1885; SCHRÖDER)

192. Kawali Sulawesi.

193. Kawali Sulawesi.

KECHUBONG
A copper helmet, imitated from comparable helmets from Europe and India. It has an almost flat top and a rounded, squared rim around it, elongated to protect ears and neck.
(GARDNER 1936)

KECHIL
A type of knife.
(STONE)

KEDOEBANG
See *geudubang*

KEDUBANG
See *geudubang*

KELABIT
See *kliau*

KELAMBU RASUL ALLAH
A sleeveless war jacket with embroidered Muslim texts rendering the wearer invulnerable.
(GARDNER 1936)

KELAVIT BOK
See *klebit bok*

KELEBIT
See *kliau*

KELEWANG
See *klewang*

KELIAU
See *kliau*

KERAMBIT
See *lawi ayam*

KERBATOE
See *akar bahar*

KERBATU
See *akar bahar*

KERIS
[CREESE, CRIS, KRIS]

The *keris* is a dagger found over a large part of the Indonesian archipelago with an almost endless variety of blades, hilts, scabbards and decoration. The *keris* always has an asymmetric double-edged blade broadening towards the hilt whereby one side protrudes more than the other. The blade may be straight or waved and is manufactured from various kinds of metal forged and mixed to form certain patterns. The scabbard has a broadened upper part.
The *keris* is not the most effective of daggers. The reason being that the point is hardly ever completely sharpened. Moreover, it is

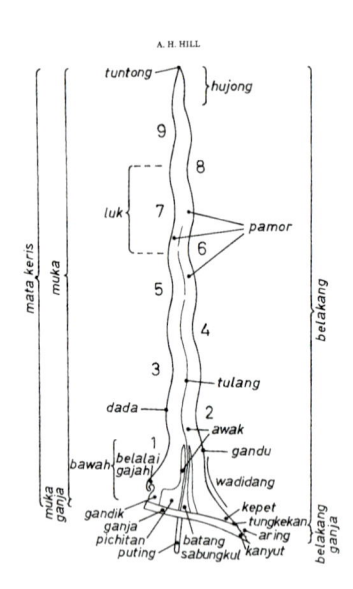

194. Keris blade
Java. Bahasa Malay terms used to describe the parts of the blade.

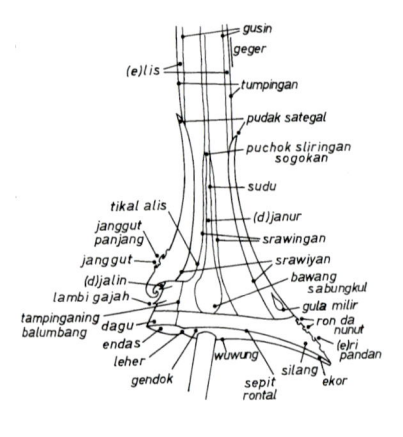

195. Keris blade
Java. Javanese terms used to describe the blade's lower parts.

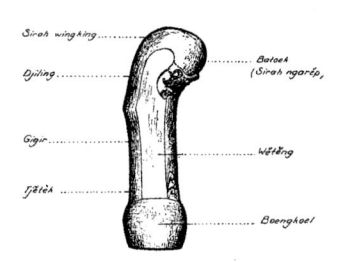

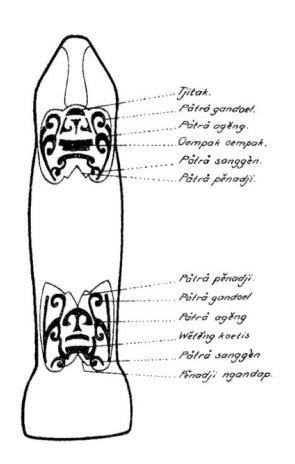

196. Keris hilt
Java, Surakarta. Names of the various details.

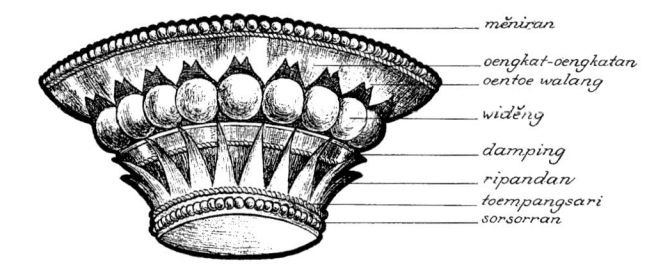

197. Keris mendak
Java. Names of the various details.

not very firm because the blade has a round prong around which a piece of cloth is wound which is then stuck into the hilt.
The following theories on the origin of the *keris* exist:
(a) it may have developed from the shape of the tip of the tail of the sting-ray (see *ikan pari*);
(b) it may have evolved from the *keris majapahit*, a small *keris* forged from a single piece of iron. The question rises if these were used as weapons, or only as luck-bringing amulets and as protectors of field crops. The term *keris majapahit* refers to the mighty Indo-Javanese empire of the 14th and 15th centuries. Its oldest examples are estimated to be over 1000 years old and must thus be much earlier than the Majapahit period. Even at present this type of *keris* is produced.
According to mythology, the *keris* was introduced on Java by the semi-divine hero Panji, the leading character in many a *wayang*. The oldest representation of a *keris* is found on a relief of the *candi* Sukuh, dating from the 14th century.
The *keris* is found over a large part of Indonesia, but by no means on all the islands. The most important areas of its distribution are Sumatra (with the exception of the Batak region), Java, Madura, Bali, Lombok, Sumbawa, Kalimantan (several coastal areas only) and Sulawesi (many coastal areas). Outside Indonesia the *keris* is found on the Malayan peninsula, the southernmost part of Thailand and southernmost islands of the Philippines.
The *keris* serves various purposes. Apart from being a defensive or an attacking weapon, it is endowed with magical powers. In the past, especially on Java, it was part of a man's daily dress and often served as a status symbol. Depending on the region or event, strict rules and regulations existed on how the *keris* had to be carried. It still plays an important part in traditional marriage ceremonies.
The *keris* is last but not least, the carrier of invisible powers. These are auspicious, protect and bring prosperity to its owner. On the other hand, evil can come from it if in the wrong hands. Countless stories are found about the supernatural powers of the *keris*. True or fantasy, they prove the belief in hidden powers is very much alive.

A *keris* blade is made by the *keris* smith, the master (*empu*), once held in great esteem and often related to the *kraton*, the palace of the sultan. The hilt and scabbard were made by other artisans. The forging was done in unassuming smithies with an open fire, a pair of bellows, an anvil and simple tools. The work included many rituals. Offers were presented, prayers said. One was aware of auspicious and inauspicious days to begin working on a *keris*. The smith was sometimes dressed in white. All this to see to it that the product would protect its future owner, bringing him prosperity and good luck.

The shape of a *keris* blade, determined by its length, width, number of undulations and the diverse variations of its base, has three main characteristics:

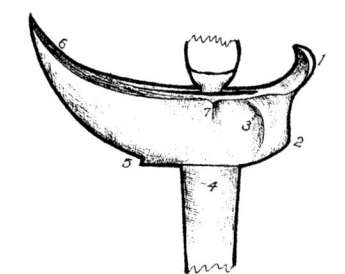

1. Angkoep
2. Djanggoet
3. Låtå
4. Gandar (in Solo antoep-antoepan).
5. Gandèk („ „ ri pandan)
6. Godong
7. Ri tjangkring.

198. Keris wranka
Java. *Wrangka brangga* or *ladrang*.

(a) its shape, either straight (*dapur bener*) or undulating (*dapur luk*). The former symbolises the mythical snake (*naga*) at rest, the latter its movements (always with an uneven number of undulations varying from three to twenty-nine or more);
(b) its details (*prabot*) such as elevated or lower parts, incrustations, curls, protrusions and small teeth;
(c) its structure (*pamor*) brought into being by forging together various kinds of metal to form patterns. Next, citric acid is applied to it. Softer types of metal are corroded causing a relief. If one metal contains nickel while the other does not, the adding of arsenic (*waringin*) causes the metal without nickel to turn black, while the nickel-containing metal remains as shiny as silver. This results in beautiful patterns in the metal. On the blade's lower part we see the *ganja* forming the separate base. The *ganja*'s protruding segment may have upright teeth or spikes (*greneng*).

The hilt (*ukiran*) of the *keris* has two or sometimes three segments:
(a) the hilt itself, which may be made of a large variety of materials in an enormous variation of forms, renditions and decorations;
(b) a shaft ring (*mendak*) located between the hilt itself and the blade (*wilah*);
(c) a bowl-shaped ornament (*selut*) sometimes encloses the lower part of the hilt.

199. Keris pendok
Java. Three different types.

64
TRADITIONAL WEAPONS OF THE INDONESIAN ARCHIPELAGO

200. Keris
Sumatra, Minangkabau. L. 35 cm.

201. Keris
Sumatra, Minangkabau (?). L. 32 cm.

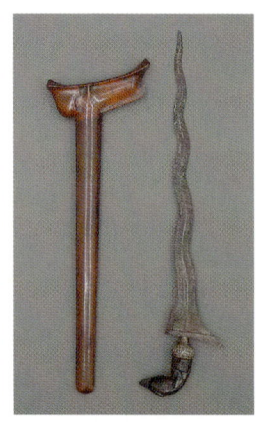

202. Keris
South Sumatra, Palembang. L. 47 cm.

203. Keris
Sumatra, Palembang. Hilt: *jawa demam*. L. 42.5 cm.

204. Keris
Sumatra, Minangkabau. L. 37 cm.

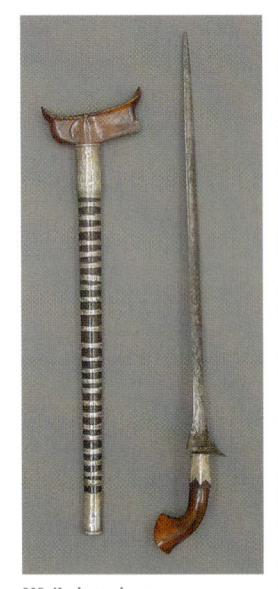

205. Keris panjang
Sumatra. L. 57.5 cm.

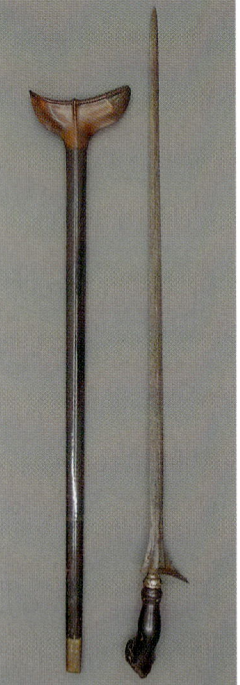

206. Keris panjang
Sumatra. L. 73 cm.

207. Keris panjang
Sumatra. L. 67 cm.

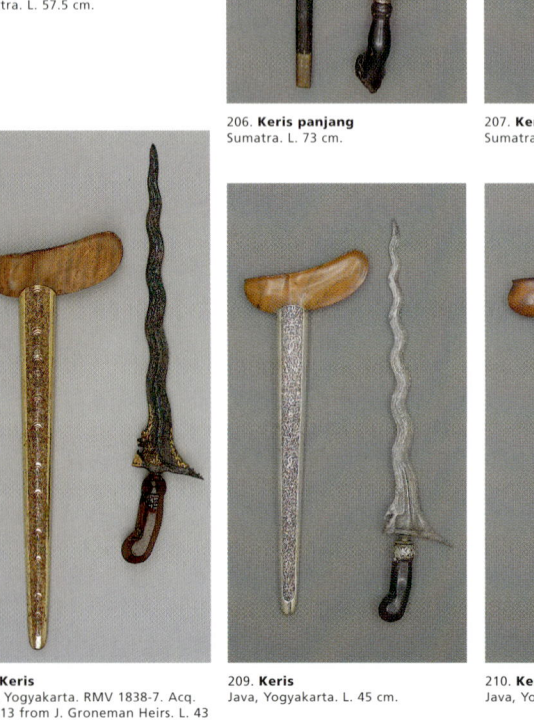

208. Keris
Java, Yogyakarta. RMV 1838-7. Acq. in 1913 from J. Groneman Heirs. L. 43 cm.

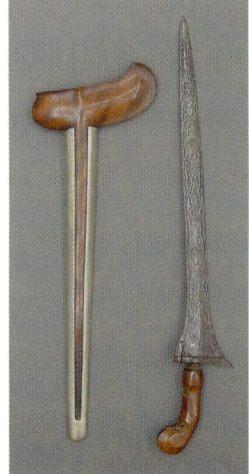

209. Keris
Java, Yogyakarta. L. 45 cm.

210. Keris
Java, Yogyakarta. L. 51 cm.

211. Keris
Java, Yogyakarta. RMV 360-1481a. Acq. in 1883 through the Koninklijk Kabinet van Zeldzaamheden, Den Haag. This *keris* was once owned by the Sultan of Yogyakarta. L. 53 cm.

ALPHABETICAL SURVEY

212. Keris
Java, Surakarta. L. 46.5 cm.

213. Keris
Java, Surakarta. L. 46 cm.

214. Keris
Java / Sulawesi. RMV 360-6021. Acq. in 1883 through the Koninklijk Kabinet van Zeldzaamheden, Den Haag. Collected before 1750. L. 46 cm.

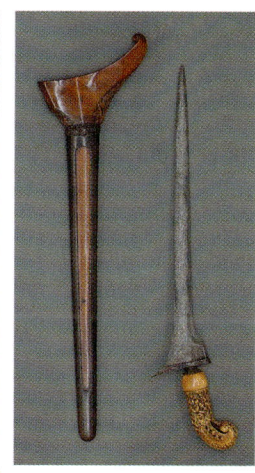

215. Keris
East Java / Madura. L. 41 cm.

216. Keris
Bali.

217. Keris
Bali. L. 53 cm.

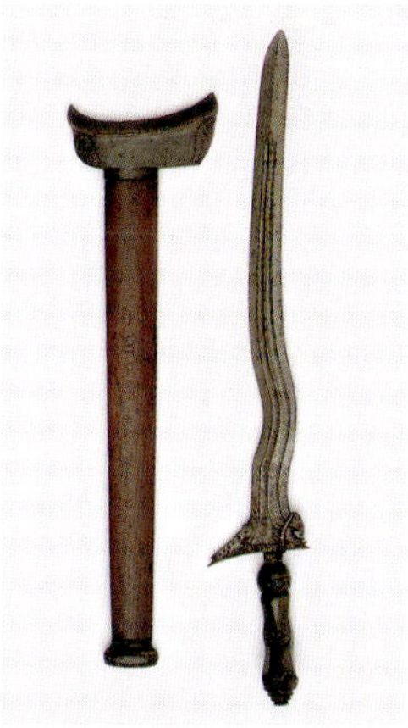

218. Keris
Bali. RMV 3600-193. Acq. in 1959 through the Ethnografisch Museum van de Koninklijke Militaire Academie, Breda. L. 53 cm.

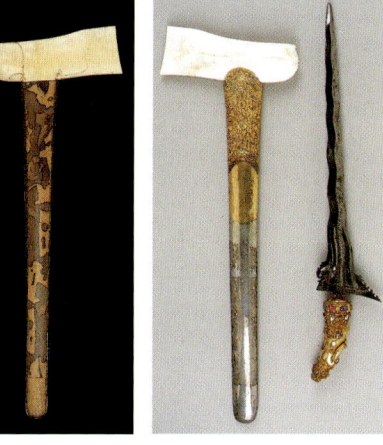

219. Keris
Rear view of illus. 218.

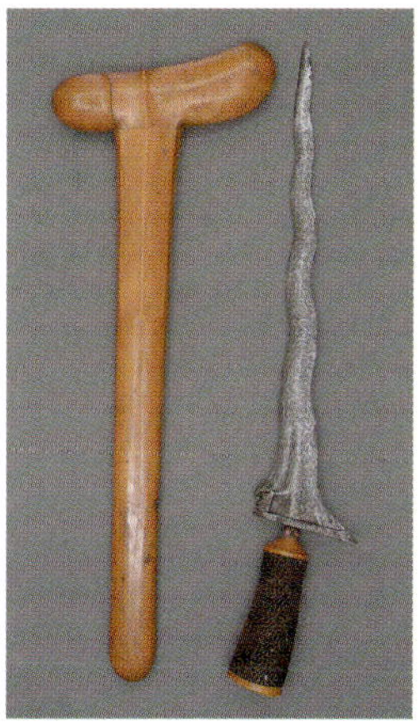

220. Keris
Bali. L. 61.5 cm.

221. Keris
Kalimantan. RMV 970-23. Acq. in 1893. L. 66.5 cm.

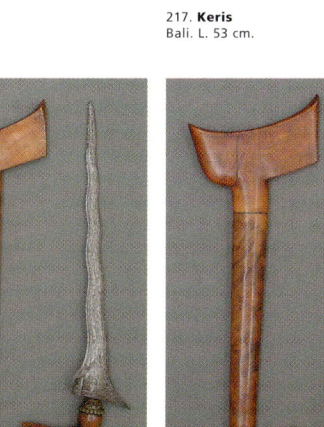

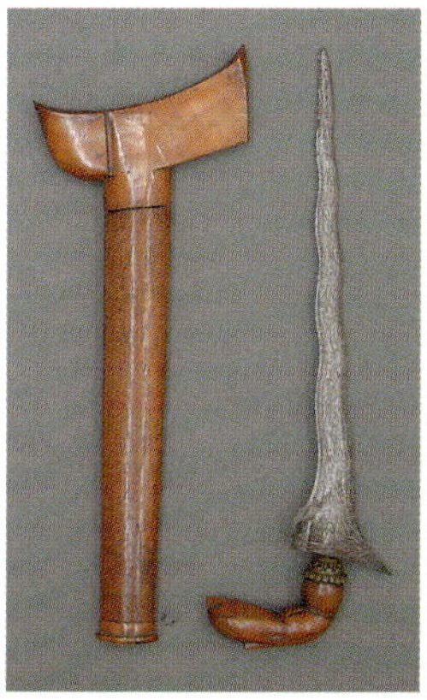

222. Keris
South Sulawesi, Buginese. L. 42.5 cm.

223. Keris
South Sulawesi, Buginese. L. 42.5 cm.

224. Keris
Sulawesi, Buginese. RMV. 1522-1 Acq. in 1906 from the widow Lansberge. L. 44 cm.

TRADITIONAL WEAPONS OF THE INDONESIAN ARCHIPELAGO

225. Keris hilt, Jawa Demam
Central Sumatra (Minangkabau ?). *Tridacna* (clam shell). H. 8.5 cm.

226. Keris hilt, Jawa Demam
Central Sumatra (Minangkabau ?). Fossil molar of an elephant. H. 7.8 cm.

227. Keris hilt, Jawa Demam
Central Sumatra. Ivory, brass. H. 7.9 cm.

228. Keris hilt, Jawa Demam
South Sumatra, Palembang. Ivory. H. 6.5 cm.

229. Keris hilt
South Sumatra, Lampung. Wood. H. 9 cm.

230. Keris hilt
South Sumatra, Lampung. Wood, silver. H. 9 cm.

231. Keris hilt
South Sumatra, Palembang (also occurs on Java, Surakarta). Wood. H. 8 cm.

232. Keris hilt
North east Sumatra, Karimun islands. Wood, silver. H. 7.6 cm.

233. Keris hilt
North east Sumatra, Karimun islands. Ivory. H. 10.5 cm.

234. Keris hilt
Java, Yogyakarta. Ivory. H. 7.5 cm.

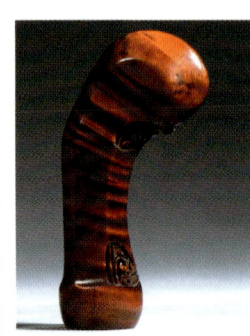

235. Keris hilt
Java, Surakarta. Wood. H. 9 cm.

236. Keris hilt
Java (Madura ?). Wood (*pelet*). H. 9 cm.

237. Keris hilt, Ganesha
Java, Cirebon. Wood. H. 9.5 cm.

ALPHABETICAL SURVEY

238. Keris hilt, Ancestor
Java, north coast, Cirebon. Wood, metal with traces of gilt. H. 9 cm.

239. Keris hilt, Rajamala (Wayang figure)
Java, Tegal. Wood, silver. H. 12.5 cm.

240. Keris hilt, Raksasa
Java, north coast. Ivory, silver. H. 8.8 cm.

241. Keris hilt, Kala (boar ?)
Java. Wood, traces of polychrome paint. H. 9.8 cm.

242. Keris hilt
Java, Madura. Ivory. H. 9 cm.

243. Keris hilt
Rear view of illus. 242.

244. Keris hilt, Tree of Life (Gunungan)
Java, Madura. Ivory. H. 9.8 cm.

245. Keris hilt, Cockatoo
Java, Madura. Ivory. H. 8.5 cm.

246. Keris hilt, Cecekahan
Bali. Polychrome wood. H. 11 cm.

247. Keris hilt, Cecanginan
Bali. Polychrome wood, ivory. H. 11.5 cm.

248. Keris hilt, Gerantiman
Bali. Silver. H. 12.5 cm.

249. Keris hilt, Kocet-Kocetan
Bali. Wood, copper, glass. H. 10.2 cm.

250. Keris hilt, Bima (Wayang figure)
Bali. Brass, glass. H. 11.7 cm.

251. Keris hilt, Buta
Bali. Wood. H. 11.6 cm.

252. Keris hilt
South Sulawesi, Buginese. Ivory.
H. 5.7 cm.

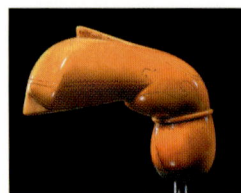

253. Keris hilt
South Sulawesi, Buginese. Ivory.
H. 6.5 cm.

254. Keris hilt
South Sulawesi, collected on
Sumbawa. Silver. H. 8.3 cm.

255. Keris hilt
South Kalimantan, collected in
Banjarmasin. Bronze. H. 7.3 cm.

256. Keris hilt
South Kalimantan, Banjarmasin. Gilt brass, *intan* (diamond). H. 6.5 cm.

Occasionally no *mendak* is found, only a *selut*.
The scabbard also has a large variety of renditions, usually determined by region. Its parts are:
(a) a broad upper segment (*wrangka* or *wrongka*) in which the blade's broadened base rests;
(b) a slender, tube-shaped lower segment enclosing the blade's narrower part called *gandar* which may have a metal scabbard (*pendok*) which completely covers the *gandar*. Otherwise we find on the front side lengthwise an opening through which we can see the wood of the *gandar*. This opening may be filled up with a thin piece of decorated metal or with turtle shell.

(VAN DUUREN 1996A, 1996B; FREY; GARDNER 1936; GRONEMAN 1910A; HARSRINUKSMO; HILL 1970; HUYSER 1916; JASPER AND PIRNGADIE 1930; JENSEN; KERNER; SOLYOM; TAMMENS)

KERIS MAJAPAHIT
JAVA

An ancient, small *keris-cum*-amulet. Its hilt and blade are entirely made of iron and forged from a single piece. This type of *keris* has often been passed down from father to son through the generations as time honoured *pusakas*, family heirlooms. They are said to be potent amulets which protect the crops in the field against disease and vermin. The name *keris majapahit* originates from the mighty Indo-Javanese empire called Majapahit, which spread across the Indonesian archipelago in the 14th and 15th century. However, this *keris* dates from much earlier times. According to estimations the first examples were made 1000-1600 years ago, the earliest Iron Age on Java. They were produced during a long period, even till recently. Later examples increasingly resemble 'normal' *kerisses*. The blade is rather roughly forged. The asymmetrical broadening at its base is far less pronounced than that of the normal *keris*. The *keris majapahit* does not have a forged *ganja*, but sometimes these are indicated by scratching in the blade. Most blades are straight, but undulating blades also occur. On nearly all examples of such *kerisses* traces of welded strips of iron can be found. Simple *pamor* is common.

The hilt has the shape of a more or less stylised human figure, sometimes standing, sometimes squatting. This figurine often holds its arms folded across the chest and has a kind of small cap on its head. Simpler examples have only a smooth body and head. The front of the figurine usually faces the flat side of the blade. Some examples show the figurine facing the edge (on the side where the base of the blade protrudes least).

A *keris majapahit* was presumably never used as a stabbing weapon. It usually goes without a scabbard. If it does have one, the scabbard is of a more recent date, and often very plain.

(BAKAR BIN PAWANCHEE; DRAEGER; VAN DUUREN 1996A; FREY; GARDNER 1933, 1936; HAZEU; HILL 1970; HOOP 1949; KERNER; MUSEUM 1965; SCHMELTZ 1890; SOLYOM)

257. **Keris majapahit**
Java. RMV 913-75. Acq. in 1892 from dr. J. Groneman. L. 27 cm.

KERODJON
See *bawar*

KEROJON
See *bawar*

KETOEPONG
See *ketupung*

KETUPUNG
[KETOEPONG, TENGKOELOE, TENGKULU]

SUMATRA, ACEH

A combat head-dress made of buffalo hide or quilted cotton leaving the face uncovered.
(KREEMER)

KIAM BOKIAM
JAVA, MADURA

A straight sword used in the martial art called *kuntao*.
(DRAEGER)

KISA
BANDA, KISSER PEOPLE

A scabbard of the *raa*.
(FOY)

KLAMBI TAYAH
[KLAMBI TAIAH, BAJU TILAM]

KALIMANTAN, SEA DAYAK

A quilted jacket, part of a combat dress. The jacket is made of cotton, usually without sleeves or a collar. The striped version is most common. This jacket is thick enough to protect against javelins only. A shield is, therefore, used against other weapons.
(ROTH 1896B)

KLANGKAH
See *sikin pasangan*

KLAU
See *kliau*

KLAUBUK
EAST KALIMANTAN

A type of shield.
(HEIN 1890)

KLAWANG
See *kliau*

KLEBIT
See *kliau*

258. **Klebit bok**
Kalimantan. L. 115 cm.

259. **Klebit bok**
Rear view of illus. 258.

KLEBIT BOK
[KELAVIT BOK]

KALIMANTAN, BAHAU, KAYAN, KENYAH

A shield in the shape of a *kliau* and painted on both sides. The front is richly decorated with clumps of human hair forcefully pressed into the narrow cracks of wood before being secured by means of fresh wax. The hair is allegedly from hunted heads. It completes the designs, making the shield all the more terrifying

260. Klebit bok
Kalimantan. RMV 1525-2. Acq. in 1906 from Lansberge Heirs. L. 130 cm, W. 44.5 cm.

for one's enemies. Its name is derived from *klebit* (meaning: 'shield') and *bok* (meaning: 'hair') indicating a strong link between the shield and the hair with which it is decorated.
(BODROGI; FURNESS; HOSE)

KLEVANG
See *klewang*

KLEWANG
[GLIWANG, KALEWANG, KELEWANG, KLEVANG, LAMANG, LAMENG]

A collective noun for swords and machetes of various shapes and origin found all over the Indonesian archipelago A characteristic of the *klewang* is that the blade widens towards the point, so that the centre of gravity lies near the tip. The blade is straight or slightly curved.

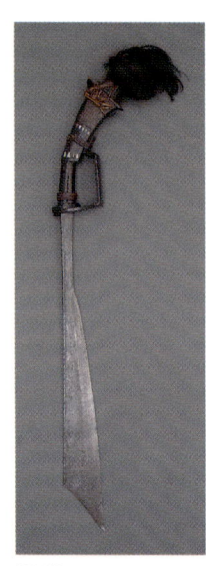

261. Klewang
Maluku. L. 80.5 cm.

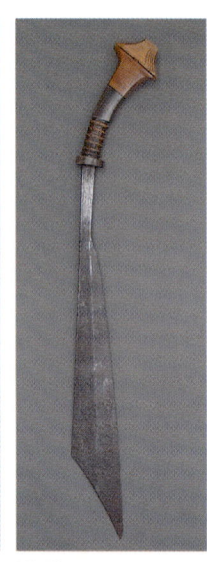

262. Klewang
Sulawesi (?). L. 83 cm.

263. Klewang
Sumatra, Palembang. L. 86.5 cm.

264. Klewang
Sumatra. Hilt: *hulu tapa guda*. L. 60.5 cm.

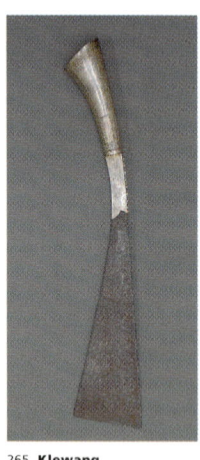

265. Klewang
South Sumatra. L. 50 cm.

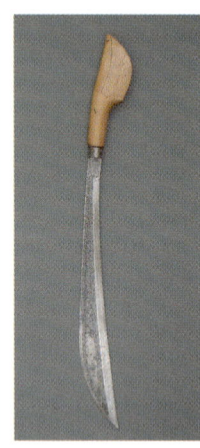

266. Klewang
Sumatra, Padang Highlands. L. 57.5 cm.

267. Klewang
Sumatra. L. 52.5 cm.

268. Klewang
Sumatra, Batak. Hilt: *hulu iku mie*. L. 68 cm.

ALPHABETICAL SURVEY

269. Klewang
Sumatra, Batak. L. 52 cm.

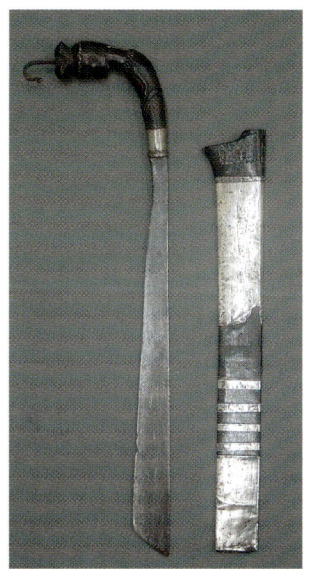

270. Klewang
Buton. RMV 781-103. Acq. in 1890 from W.E.M.S. Aernout. L. 85 cm.

271. Klewang
Leti. Warrior carrying a sword (*klewang*).

272. Klewang
Timor. RMV 360-5584. Acq. in 1883 through the Koninklijk Kabinet van Zeldzaamheden, Den Haag. L. 113 cm.

KLEWANG BELADAN
See *barong*

KLEWANG CARA ACEH
[KLEWANG TJARA ACEH]
Sumatra, Aceh
A certain type of *klewang*.
(STONE)

KLEWANG COK JANG
See *co jang*

KLEWANG LAMTRIENG
See *rudus*

KLEWANG LANGTRIENG
See *rudus*

KLEWANG LIPEH OEDJONG
See *rudus*

KLEWANG LIPEH UJONG
See *rudus*

KLEWANG PEH LAM TRIENG
See *rudus*

KLEWANG POETJOK MEUKAWET
See *klewang pucok meukawet*

KLEWANG PUCHOK BERKAIT
See *balato*

KLEWANG PUCOK MEUKAWET
[KLEWANG POETJOK MEUKAWET, KLEWANG PUTJOK MOKAWET]
SUMATRA
A variant of the *klewang*. The blade has a straight edge for most of its length. Near the tip, however, we see an almost right-angled curve in the direction of the edge. The back is also straight and bends at the tip in a sharp curve to the point.
(ROGERS; VERMEIREN 1983)

KLEWANG PUTJOK MOKAWET
See *klewang pucok meukawet*

KLEWANG TARAK BAGOE
See *tarah baju*

KLEWANG TEBAJ OEDJONG
See *klewang tebal hujong*

KLEWANG TEBAL HUJONG
[GOROPOH PASE, KLEWANG TEBAJ OEDJONG, KLEWANG TEBAL OEDJOENG, KLEWANG TEBAL OEDJONG, KLEWANG TEUBAJ OEDJONG, KLEWANG TOBAJ UDJONG]
SUMATRA, ACEH
A *klewang* with an almost straight edge and back with a thick, broadly tapering point. The hilt is made of horn, is slightly curved and may have a sharp protrusion at the top.
(DIELES; JACOBS; NIEUWENHUIZEN; ROGERS; SNOUCK HURGRONJE 1904)

KLEWANG TEBAL OEDJOENG
See *klewang tebal hujong*

KLEWANG TEBAL OEDJONG
See *klewang tebal hujong*

KLEWANG TEUBAJ OEDJONG
See *klewang tebal hujong*

KLEWANG TJARA ACEH
See *klewang cara aceh*

KLEWANG TJOK JANG
See *co jang*

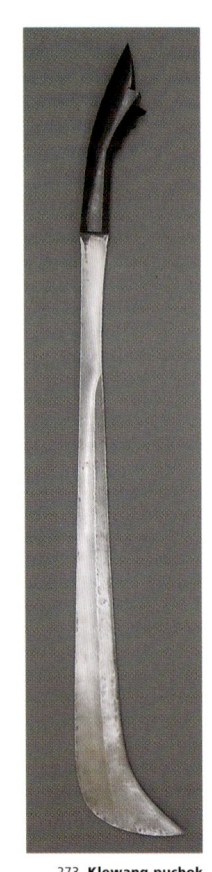

KLEWANG TJOT JANG
See *co jang*

KLEWANG TOBAJ UDJONG
See *klewang tebal hujong*

273. **Klewang puchok meukawet**
Sumatra, Aceh. Hilt: *hulu tapa guda*. L. 78 cm.

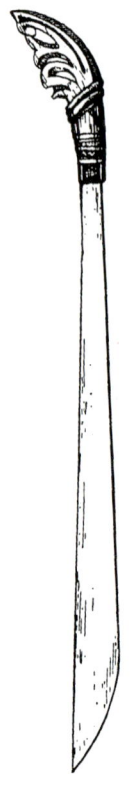

274. **Klewang tebal hujong**
North Sumatra Hilt: *hulu iku ite*.

275. **Kliau**
Kalimantan. RMV 405-57. Acq. in 1883 from the Sultan of Kutei. L. 134.5 cm, W. 39.5 cm.

of intricate figures on the inside and the outside. Amongst some groups the shield is totally or partially covered with tufts of human hair taken from the warrior's victims (see *klebit bok*). In addition, *nassa* shells and mother-of-pearl can be glued onto it with resin as a means of decoration. Elements such as fanciful dragon heads or faces, connected by means of *aso*-figures, may also serve as ornament. Moreover, human figures, hornbills and geometric motifs are found.

This shield, held at a certain distance from the body, scares off the enemy and, of course, protects the person carrying it.

(FELDMAN 1985; GOMES; HEIN 1890; HEIN 1899; HOSE; ROTH 1896B; SELLATO; STONE).

276. Kliau
Kalimantan. RMV 3600-402. Acq. in 1959 through the Ethnografisch Museum van de Koninklijke Militaire Academie, Breda. L. 130 cm, W. 42 cm.

277. Kliau
Kalimantan. L. 131 cm.

278. Kliau
Rear view of illus. 277.

279. Kliau
Kalimantan. L. 115 cm.

280. Kliau
Rear view of illus. 279.

KLIAU
[KELEBIT, KELIAU, KLAU, KLAWANG, KLEBIT, TRABAI]

KALIMANTAN, SOUTH SULAWESI, CENTRAL SULAWESI

The most popular shield, especially amongst the Dayak. The *kliau* and its massive grip are both usually carved from a single piece of soft, light *jelutong* wood. This grip is located lengthwise in the centre of the shield. The length of the shield varies from c.90-100 cm, the width from c.40-50 cm. Its sides are almost straight and parallel. Both its tips end in a point. It is slightly curved both in the width and lengthwise. In the centre we see a slight angle along the entire length. The *kliau* is usually strengthened in the width with rattan strips to prevent the wood splitting when struck by a sword.

The decoration varies from having a simple painting on the ends to being entirely painted in one or more colours, showing all kinds

TRADITIONAL WEAPONS OF THE INDONESIAN ARCHIPELAGO

KNIFE

A large variety of knives is found all over the Indonesian archipelago. Their specific names remain unknown. A number of them is represented below.

281. Knife
Sumbawa.

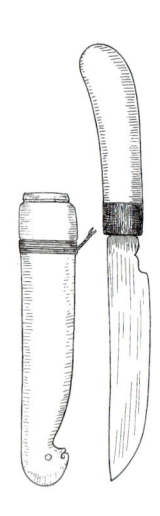

282. Knife
Sumbawa.

283. Knife
Sumbawa, Bima. L. 27.5 cm.

284. Knife
West Java. L. 34 cm.

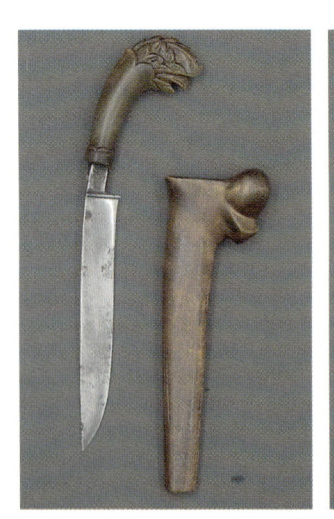

285. Knife
Sumatra. L. 20 cm.

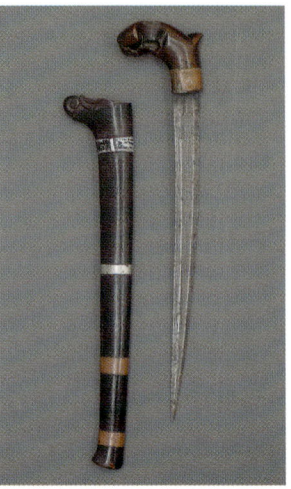

286. Knife
Salayer. L. 40 cm.

287. Ceremonial knife
Bali, Bulelang. RMV 1050-2. Acq. in 1895 from mr. L.A.J.W. Sloet Van Der Beele Van Nispen Heirs. L. 45.5 cm.

KODHIK
See *kudi*

KOEDI
See *kudi*

KOEDJANG
See *kudi*

KOELANGKAH
See *sikin pasangan*

KOELANGKAIH
See *sikin pasangan*

KOENJOER
See *leumbeng*

KOERAMBI
See *lawi ayam*

KOERAMBIT
See *lawi ayam*

KOETA DJA
See *kuta ja*

KOHONG KALUNAN
KALIMANTAN, DAYAK

A pattern of fanciful faces as used to decorate the hilts of *mandau*s.
(STONE)

KORAIBI
See *kurabit*

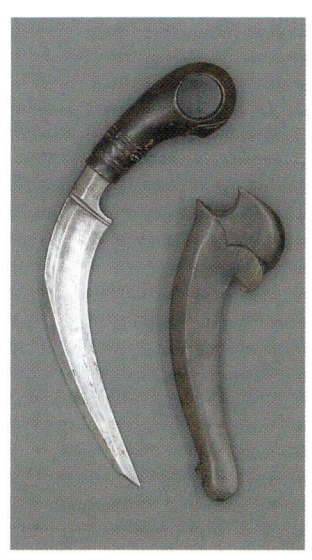

288. **Korambi**
Sumatra. L. 19 cm.

289. **Korambi**
Sumatra. L. 19 cm.

KORAMBI
[PISO KOERAMBI, PISO KURAMBI]

SUMATRA, SULAWESI

A knife with a short, crescent shaped, strong blade. The edge is located on the inside of the blade. Part of the back is often sharpened, too. Lengthwise one or more grooves may be seen. The hilt may have all types of forms. It is usually thickened at the end and has a hole through which to place the index finger forming an additional support. The scabbard follows the blade's shape and is often partially open at the rear to facilitate the withdrawal of the blade from the scabbard. The *korambi* is used with an upward jab.
(ROGERS; STONE)

KORUNG
FLORES

A name for a spear on Flores.
(DRAEGER)

KOTAU
A jacket, part of the combat dress.
(GARDNER 1936)

KOWLIUM
JAVA, MADURA

A weapon in the form of a boat-hook with a short shaft and a barbed, sharp iron spear-point.
(DRAEGER)

KRIS
See *keris*

KUDHI WAYANG
JAVA

A *kudi* with a blade in the shape of a *wayang* figure. At fixed times, usually on Thursday evening, they are objects of worship and, as are the *wayang* puppets, placed in a vase or a trunk of a banana tree. This takes place indoors only in such a way that it is impossible for outsiders to see such objects, considered sacred heirlooms, *pusaka*s.
(VAN DONGEN)

290. **Kudi wayang**
Java, Cirebon. RMV 1403-2559. Dated to about between 1600 and 1800. Acq. in 1903 through the Rijksmuseum van Oudheden, Leiden. L. 50 cm.

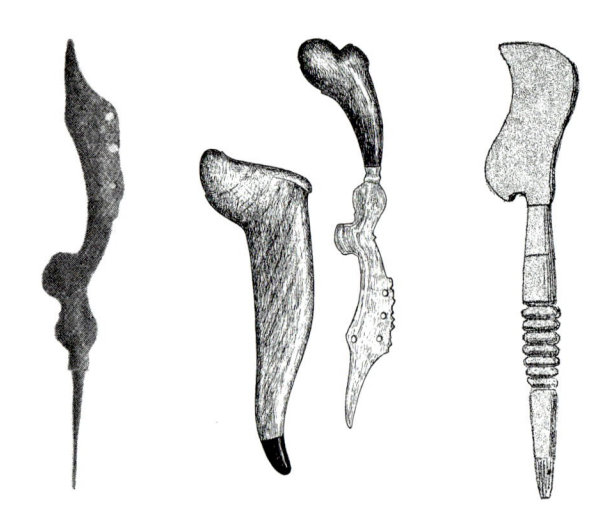

291. Kudi
Central Java. L. 16.5 cm.

292. Kudi

293. Kudi
Java.

KUDI
[BADE GAGANG BEUSI, BADI GAGANG BEUSI, KODHIK, KOEDI, KOEDJANG, KUDIK, TJUNDRIK]

JAVA, MADURA, BALI
A knife with a fancifully shaped blade, to which magical powers are attributed. The *kudi* has many shapes and sizes. The oldest examples are small and date from a time that iron was rare. They share a great similarity with the *keris majapahit*. The *kudi*'s one-sided blade and its hilt are forged from one piece. The latter often has the form of a human or human head facing the edge. *Kudi*s usually served as a talisman and were carried, for instance, by a medicine man (*pawang*) as a symbol of his power.
Although this type of *kudi* is ancient, one continued to make them. Later models are usually larger, varying from the format of a small knife to the size of a machete. Sometimes they resemble axes and have a long round wooden or horn shaft as a hilt (*garan*). The *kudi* may also be placed upon a shaft of a spear.
The blade has a large variety of forms in which sometimes the shape of a bird's head (*kudi peksi*), a snake's head (*kudi naga antaboga*) a monkey's head (*kudi bedes*), a dragon or *wayang* figure (*kudi wayang*) can be recognised. Blades often have one or more small holes, sometimes filled with copper. The *kudi* may or may not have a scabbard
(DRAEGER; GARDNER 1936; HAZEU; JASPER AND PIRNGADIE 1930; KREEMER; MUSEUM 1965; PARAVICINI; RAFFLES 1817A; SNOUCK HURGRONJE 1892; STONE)

KUDI TRANCHANG
A knife with a fancifully shaped blade and a long hilt, to which magical powers are attributed.
(PARAVICINI)

294. Kudi tranchang
L. 92 cm.

295. Kudi tranchang
Bali. L. 59.5 cm.

KUDIK
See *kudi*

KUJUNGI
JAVA, MADURA
A knife with a mystical significance. The fanciful blade has a number of protrusions.
(DRAEGER)

KUJUR
[KUNJUR]

SUMATRA, KUBU
A spear, still in use, mainly for hunting, with a c.2.5 m long shaft and iron tip. It is not thrown over the shoulder but thrust underhand. While one hand pushes this weapon, the other guides it. Prey is thus pierced at close range. The Kubu of south Sumatra use various spear tips such as a barbed, triangular one for monkeys, a lancet shaped one for pigs and deer, and a barbed three-pointed one for turtles. The spear-points are acquired through exchange. The *kujur* is never decorated.
(DRAEGER; GARDNER 1936; DE JONGE 1990; MARSDEN)

KUKU AYAM
See *lawi ayam*

KUNJUR
See *kujur*

KURABIT
KURAIBI]

MENTAWAI ISLANDS
A long, slightly curved wooden shield. Its straight upper side is rather broad. Looking down it, the shield becomes a little narrower and then somewhat broader until about halfway. From here it tapers towards a blunt tip. Just above the centre the grip is cut out of the shield. The subsequent holes are covered using half a coconut shell attached by means of rattan. A *kurabit* is usually decorated with geometrical, coloured motifs on both sides. Spirals (*patogalik*) are much used. Moreover, we see decorations such as lizard and human figures in 'rock painting style'. To make red paint, the fruit of the *kalumangan* tree is used. The black paint consists of soot mixed with plant juices.
Since the termination of head-hunting and of disputes between Mentawai ethnic groups at the

296. Kurabit
Mentawai, Siberut, Katorei. RMV 79-1.
Painted in red and black. L. 115 cm, W. 28 cm.

beginning of the 20th century, the *kurabit* has become vary rare. The older examples seem to be less broad and less curved than more recent ones.
(FISCHER 1909; MUSEUM 1965; STONE; VOSKUIL 1921; WIGGERS)

KUTA JA
[KOETA DJA]

SUMATRA, ACEH

A shield named *kuta ja* (meaning: 'walking fortress') made of a semi-cylindrically curved piece of buffalo hide. It could serve as a protection for a number of people at the same time. Indeed, during inland disputes, warriors sometimes advanced in groups hiding behind a *kuta ja*, off which bullets were said to ricochet.
(KREEMER)

KWAI

JAVA, MADURA

A bludgeon-shaped weapon consisting of a long shaft with a protrusion attached cross-wise.
(DRAEGER)

297. **Kurabit**
Mentawai. RMV 3600-1200. Acq. in 1923. L. 110 cm, W. 36 cm.

298. **Lopah petawaran**
Sumatra, Batak. RMV 3155-20. Acq. in 1954 from W.G. Broek. L. 30 cm.

ALPHABETICAL SURVEY

LABE
FLORES
A type of arrow.
(DRAEGER)

LABO
[PADE, TOLAKI]
SULAWESI, TORAJA
A general term for a machete. The name *tolaki* is used in (the vicinity of) Kendari Bay.
(DRAEGER)

LABO BALANGE
SULAWESI
A sword meant for use in time of war.
(DRAEGER)

LABO BALE BALE
SULAWESI
A knife used to slaughter buffaloes and as a weapon in case of emergency.
(DRAEGER)

LABO TO DOLO
SULAWESI, TORAJA
An 'ancestral' sword.
(RODGERS)

LABO TOPANG
SULAWESI
A machete also used as a weapon.
(DRAEGER)

LADEENG
See *ladieng*

LADENG
See *ladieng*

LADIENG
[LADEENG, LADENG, ROEDOES LENTI, ROEDOES LENTIK, RUDUS LENTI, SONAGANG-KLEWANG]
SUMATRA, ACEH, GAYO, ALAS
A sword, once a formidable combat weapon. At the beginning of the 20th century, however, it was used as a machete and a knife preferred when slaying sacrificial animals. It is held in the right hand and goes without a scabbard. The blade's back is concave and curves towards the convex edge at the tip. The blade which broadens towards the tip sometimes shows beautiful lines thanks to the mixed forging of iron and steel. After exposing the blades to the caustic effect of lemon juice, these lines are accentuated. Sometimes sulphuric arsenic (*warang* or *tuba tikoih*; Gayo: *jenu* or *tubo tikus*) is added to this juice resulting in the dark, almost black colour of iron arsenic. The people of Aceh call these lines *kuree*. In Gayo and Alas they are known as *kure*. The name *ladieng* is used in Aceh, while *rudus lenti* is found in Gayo and Alas. The hilt is an example of the *hulu tapa guda* type.
(JACOBS; KREEMER; VAN LANGEN; NIEUWENHUIZEN; ROGERS; SNOUCK HURGRONJE 1892)

LADIING BOENGKOEWQ
See *sadeueb*

LADIING PARAMBAH
See *sadeueb*

299. **Ladieng**
Sumatra, Aceh. Hilt: *hulu tapa guda*. L. 76 cm.

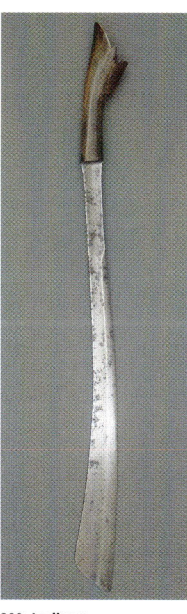

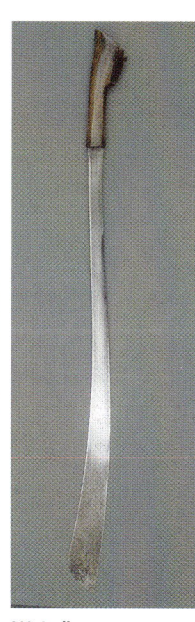

300. **Ladieng**
Sumatra, Aceh. Hilt: *hulu tapa guda*. L. 72.5 cm.

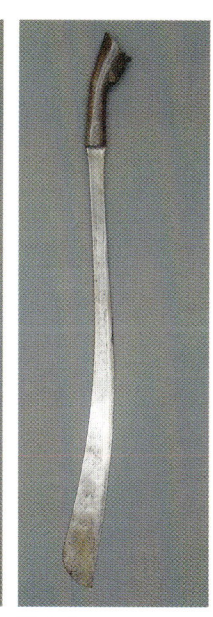

301. **Ladieng**
Sumatra, Aceh. Hilt: *hulu tapa guda*. L. 72 cm.

LADING
See *sikin pasangan*

LADING BELAJOENG LAMAH
See *lading belajung lamah*

LADING BELAJUNG LAMAH
[LADING BELAJOENG LAMAH]
KALIMANTAN
A sabre with a short, heavy blade broadening towards the point. The hilt is in the shape of a pistol-grip leaving the hand unprotected.
(STONE)

LADING CARA
[LADING TJARA]
KALIMANTAN, DAYAK
A knife 'with an Arabian form'.
(STONE)

LADING DJAWA
See *lading Jawa*

LADING JAWA
[LADING DJAWA]
KALIMANTAN, DAYAK
A knife 'with a Javanese form'.
(STONE)

LADING TERUS
A dagger with a blade resembling the tip of a spear or arrow. The hilt often resembles the hilt of the *lawi ayam* with a semi-circular indentation on top. The weapon is used underhand and may be derived from a short javelin.
(GARDNER 1936)

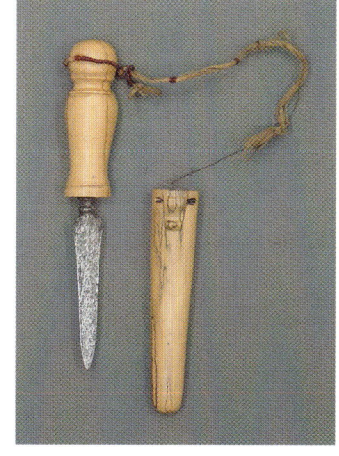

302. **Lading terus**
L. 16 cm.

LADING TJARA
See *lading cara*

LADINGAN
See *ladingin*

LADINGIN
[LADINGAN, MERMOE PAKPAK, MERMU PAKPAK]

SUMATRA, KARO PLATEAU, TOBA BATAK, PAKPAK BATAK, GAYO-LUOS
A machete with a straight back and edge, broadening towards the tip. The sharp side ends in a curve towards the back. The hilt is usually made of buffalo horn. The tip of the hilt has two protrusions and is called *sukul ngangan* or *sukul katungangan*. However, other forms such as *hulu iku mie* (hilt 'as the knotted cat's tail') are also found. The machete has a scabbard which may have a sharp protrusion at the end.
(FISCHER 1914; FORMAN; VOLZ 1909, 1912)

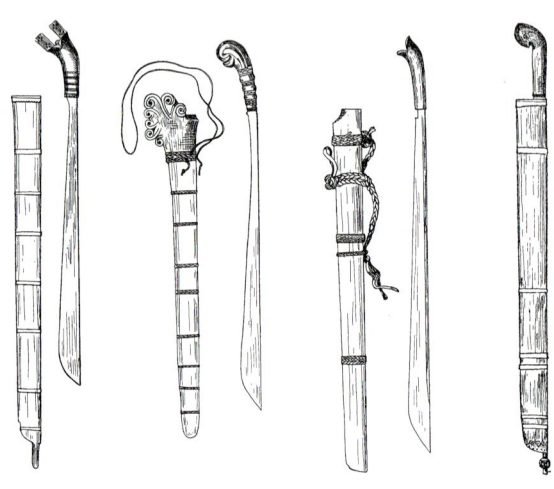

303. **Ladingin** Sumatra, Pakpak Batak. Hilt: *sukul ngangan*.
304. **Ladingin** Sumatra, Toba Batak.
305. **Ladingin** Sumatra, Toba Batak.
306. **Ladingin** Sumatra, Pakpak Batak. Hilt: *sukul jering*.

LADJAU
See *lajau*

LAJATANG
JAVA, MADURA
A weapon consisting of a stick to both ends of which crescent-shaped blades are attached. Two protrusions are found on the back of the crescents.
(DRAEGER)

LAJAU [LADJAU]
KALIMANTAN
A small arrow with a brass tip, used with a blow-pipe.
(STONE)

LAJOEK-LAJOEK
See *lajuk lajuk*

LAJUK LAJUK
[LAJOEK-LAJOEK]

SUMATRA, KARO BATAK
A sword or a knife with a broad blade and a straight, almost parallel back and edge. The edge curves at the tip in an oblique angle towards the back. The blade is narrower near the hilt and then broadens towards the edge. The hilt (made of horn or wood) is slightly curved halfway, or otherwise resembles the form of the *sukul jering*. The scabbard is straight and broadening somewhat at the mouth towards the blade's edge.
(FISCHER 1914; MÜLLER)

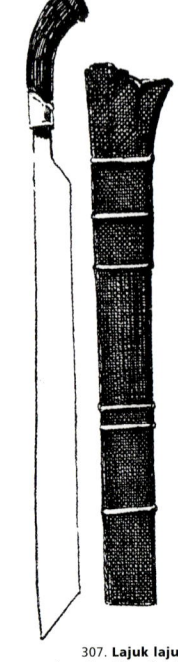

307. **Lajuk lajuk** Sumatra, Karo Batak.

LAMANG
See *klewang*

LAMBING
SUMATRA
A type of spear.
(DRAEGER; MARSDEN)

LAMENA
SULAWESI
A scaled cuirass with large copper plates. *Lamena* is both the Buginese and Makassar term. See also *baju lamina*.
(MATTHES 1874, 1885; SCHRÖDER)

LAMENG
See *klewang*

LANGA
KALIMANTAN
A dart used in a blow-pipe made of the solid fibres of several species of palm or of a light piece of bamboo. It weighs c.1 gram and measures 25 cm to more than 30 cm. Its thickest part measures c.2 mm. On its back we see a small, solid or hollow, cone made of plant marrow (of a thorny liana species) exactly fitting the blow-pipe's bore. The *langa*'s tips are usually simply straight and covered with a potent poison, mostly *ipoh*. Darts used for hunting birds are covered with poison for some 3 cm. Those shot at larger animals such as wild boars, monkeys or deer are covered with poison for some 5-6 cm. Just below the tip the dart is incised so that the tip breaks off in the wound making it difficult to remove.
Apart from the straight tip, tips with barbed hooks occur. Others are made of another material such as the *lajau* (with brass point). Darts with a tin point or a point to which small teeth of small animals of prey are glued using resin (*damar*) are also found. These darts are mainly used for hunting large wild animals.
A special instrument stored in the quiver is used to make small marrow cones. This little round stick has an outer diameter which is exactly the same as the inner diameter of the blow-pipe. In the centre we see, located lengthwise, a small pin made of metal or wood. It is used to cut the cone exactly to the size of the bore of the blow-pipe and to make the hole in which the dart is fixed. Each dart is often stored separately in a bamboo tube which fits in the quiver.
(SELLATO; STONE; WEIGLEIN)

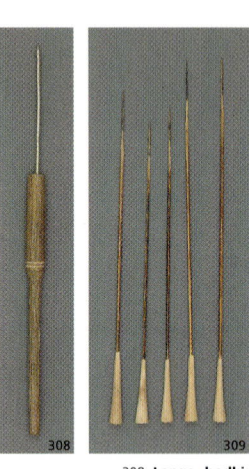

308. **Langa, bodkin** Kalimantan. Wooden bodkin with iron pin, used for making the butts of the blowpipe arrows. L. 23.5 cm.

309. **Langa** Kalimantan. L. c. 26 cm.

310. **Langa bodkin** Kalimantan. Wooden bodkin with brass pin, used for making the butts of the arrows.

LANGGAI
See *piso raout*

LANGGAI TINGGAN
See *langgai tinggang*

LANGGAI TINGGANG
[LANGGAI TINGGAN, LANGGI TINGGANG, MANDAU LANGGI TINGGANG]

KALIMANTAN, SEA DAYAK
A curved sword almost identical to the *niabor*, but with a hilt resembling that of the *mandau*. The blade has a convex edge and a concave back. On both sides a broad rib runs from the finger protector to the tip. This finger protector is smaller than that of the *niabor* and further removed from the hilt. This protector is not a copy of the *kundieng* of the *niabor*, but of the *ikang* of the *mandau* which is part of a dragon.
The Sea Dayak name *crowit* (hooks) proves there is no analogy with the *kundieng*. Both sides of the blade are flat. Its upper side may have a chiselled pattern. Along the ribs one may find a plain, curling motif called *entadok* (meaning: 'caterpillar'). The scabbard, apart from being curved, is the same as that of the *mandau*. The term *langgai tinggang* means 'the longest tail feather of a hornbill' referring perhaps to the blade's rib. See also *jimpul*.
(GARDNER 1936; SHELFORD)

LANGGEI
See *piso raout*

LANGGI TINGGANG
See *langgai tinggang*

LANGKAP
BALI
A sturdy bow.
(STONE)

LAPAN SAGI
See *sikin lapan sagu*

LARBANGO
JAVA
A sabre with a faintly S-shaped blade ending in a sharp point.
(JASPER AND PIRNGADIE 1930)

LARBIDO
JAVA
A sabre with an S-shaped blade ending in a sharp point.
(JASPER AND PIRNGADIE 1930)

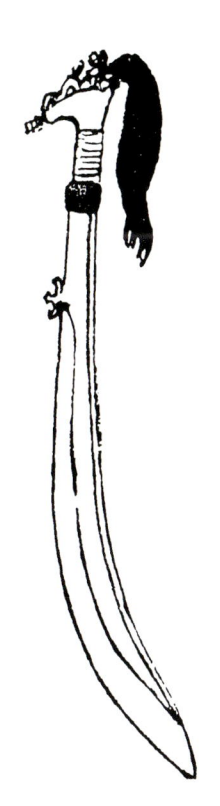

311. Langgai tinggan
Kalimantan.

312. Larbango
Java.

313. Larbido
Java.

LARKAN
MADURA
A machete with a long, decorated hilt. The blade is usually narrow, sharpened on one side and curved forwards.
(STONE)

LASAG
SULAWESI, MINDANAO (PHILIPPINES), SUBUNAN
A general term for a shield.
(STONE)

LASARA
[LASSARA]

NIAS
The *lasara* is a mythical figure linked to the god of death and darkness. In the south of Nias it takes the shape of a fanciful head with teeth of a wild bear, the mouth of a crocodile or snake, crowned by a *calao* helmet and the antlers of a deer. Decorating the outer facade by means of the heads of three *lasaras* is essential for a building where the leaders of a village in southern Nias gather. A *lasara*'s head is also added to a chief's tomb. The *balato*, the *gari* and the *tologu* have a hilt in the shape of a *lasara*'s head varying from highly stylised to fanciful ones. This head may show a human or animal figure (*bekhu*, an evil spirit) lying on its stomach, resting on its elbows and knees, and biting the *lasara*'s neck. This *bekhu* is said to appear on swords belonging to headhunting warriors and never on swords of the more peace-loving people of northern Nias. The *bekhu* is connected to the *lasara* by means of a protrusion (scruff of the neck or the tongue) sticking out of the mouth.
(BARBIER; FELDMAN 1990; DE LORM 194LA, 1942)

314. Lasara motif
Nias. Balato hilt. L. 12 cm.

315. Lasara motif
Nias. Balato hilt. L. 15 cm.

316. Lasara motif
Front view of illus. 315.

LASSARA
See *lasara*

LATOK
See *parang latok*

LATOK BUKU
See *parang latok*

LAVONG
KALIMANTAN, KAYAN
A round war cap covered with hair in various colours. On it two eyes are depicted in order to represent a face. Two long tail-feathers of the hornbill are attached on top.
(ROTH 1896B; STONE)

LAWI AYAM
[KARAMBIT, KERAMBIT, KOERAMBI, KOERAMBIT, KUKU AYAM]
SUMATRA, JAVA
A small, strongly curved crescent-shaped knife with a thin blade. The edge is located on the concave side of the blade which is sometimes sharpened on both sides. The *lawi ayam* (tail-feather of a cock) is derived from the Arabian *jambia*. This weapon is held with the thumb over the hilt's head whereby the blade points straight down and the edge to the front. It is used when stabbing upwards (*radak*) causing atrocious injuries especially amongst those dressed lightly. Women sometimes hide minute *lawi ayam*s in their hair. The hilt often has a hole in which, if large enough, the index finger can be placed. Smaller holes serve as decorations. Some hilts are finished in such a way that the hole's upper part is cut off leaving a semi-circular opening. The *lawi ayam* is carried in a scabbard or in a fold of the *sarong*.
(DRAEGER; GARDNER 1936; HILL 1970; VOSKUIL 1921)

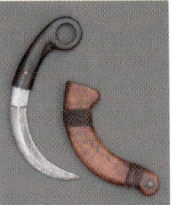

317. **Lawi ayam** L. 22 cm.

318. **Lawi ayam** Sumatra, Minangkabau. L. 17.5 cm.

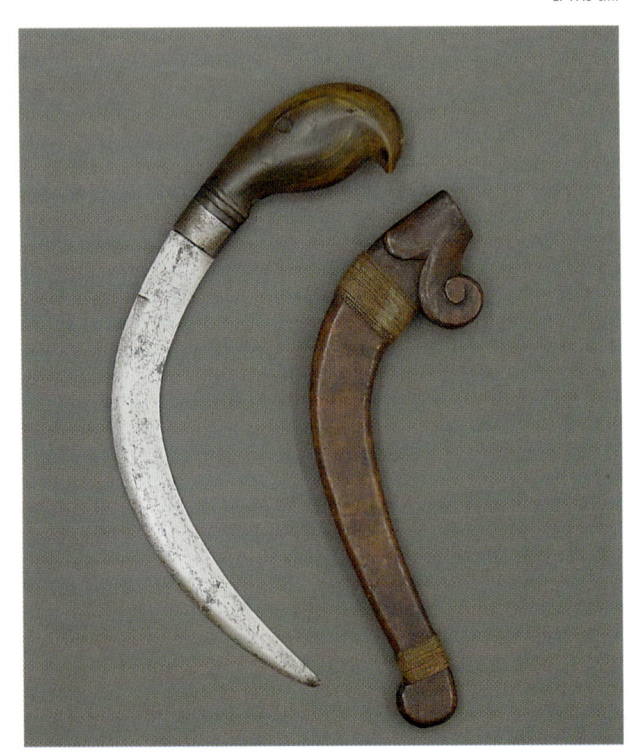

319. **Lawi ayam** L. 20 cm.

LEMBING RADJA
See *lembing raja*

LEMBING RAJA
[LEMBING RADJA]
SUMATRA, KARO BATAK
A ceremonial spear measuring c.2 m. Its very long point (c.40-45 cm) is elegantly shaped with S-shaped sides and a pointed tip. Part of the shaft has a thickened ring covered with silver plating. The *lembing raja* was only carried by dignitaries (*lembing*, meaning: 'spear'; *raja*, meaning: 'king').
(MÜLLER)

LEMBING SI DOEA-DOEA
See *lembing si dua dua*

LEMBING SI DUA DUA
[LEMBING SI-DOEA-DOEA, LEMBING SI-DUWA-DUWA]
SUMATRA, KARO BATAK
A ceremonial spear always carried in pairs (*lembing*, meaning: 'spear'; *dua*, meaning: 'two'). The length varies between c.1.8-2 m. The iron point has parallel sides, a blunt tip and a slight ridge along the middle. Such spears were mainly carried by dignitaries (*penghulu*s).
(FISCHER 1914; MÜLLER)

LEMBING SI-DUWA-DUWA
See *lembing si dua dua*

LEMING KAPAK
SUMATRA, ACEH
A spear measuring c.2 m. The tip is S-shaped on both sides, the shaft thickens near the base.
(JACOBS)

LENGOE BODONG
See *kalijawo malebu*

LENGOE-LABOE
See *kalijawo malampe*

LENGU BODONG
See *kalijawo malebu*

LENGU LABU
See *kalijawo malampe*

LEPUT
See *sumpitan*

LEUMBENG
[KAPA, KOENJOER, PENDAHAN]
SUMATRA, ACEH
A type of javelin.
(GARDNER 1936; KREEMER)

LEUNTEK
See *parang leuntek*

LIANGCAT
[LIANGTJAT]
JAVA, MADURA
A pair of sticks used in combat.
(DRAEGER)

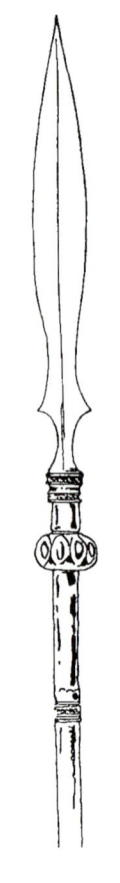

320. **Lembing raja** Sumatra, Karo Batak.

LIANGTJAT
See *liangcat*

LIDAH AYAM LIPAT
A type of knife.
(STONE)

LI PUN [PA-NA]
SULAWESI, MINDANAO (PHILIPPINES), SUBUNAN
A type of arrow.
(STONE)

LIYO LIYO
SULAWESI
The sight of the *sapu*, the Toraja blow-pipe. It is called *liyo liyo* in Buginese and Makassar.
(MATTHES 1874, 1885; SCHRÖDER)

LOEDJOE ACEH
See *sikin panjang*

LOEDJOE ALANG
See *luju alang*

LOEDJOE ALAS
See *luju alas*

LOEDJOE ATJEH
See *sikin panjang*

LOEDJOE LAPAN SAGI
See *sikin lapan sagu*

LOEDJOE MOEGENTA
see *luju mugenta*

LOEDJOE NAROE
See *sikin panjang*

LOEDJOE NARU
See *sikin panjang*

LOEDJOE TJELIKO
See *luju celiko*

LOENGKEE
See *lungkee*

LOEWOEK
See *luwuk*

LOHINGLAMBI
See *sumpitan*

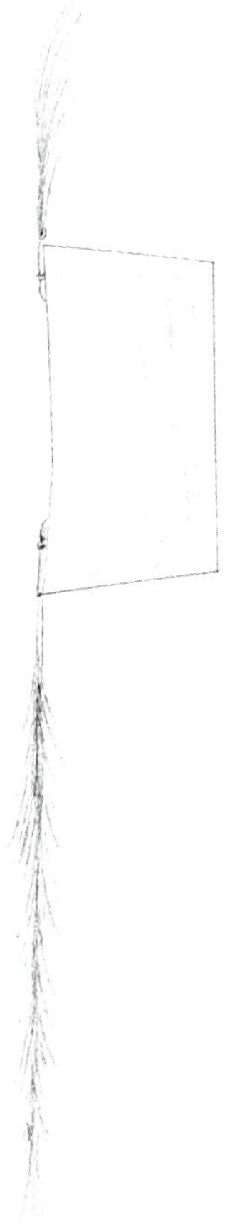

LOMBU LOMBU
[LOMBU LOMBU DJONGDJONG, LOMBU LOMBU JONGJONG]

321. **Lombu lombu**
Sumatra, Batak.

SUMATRA, BATAK
A shield made of a heavy rectangular piece of buffalo hide or wood. On the rim at the top we see a 'crest' and on the lower part a 'tail'. All this to imitate an ox (*lombu*). As the oxen of the Toba Batak are black, the colour of the *lombu lombu* is black, too. The term *lombu lombu jongjong* means: 'standing ox-representation'.
(STONE; VAN DER TUUK)

LOMBU LOMBU DJONGDJONG
See *lombu lombu*

LOMBU LOMBU JONGJONG
See *lombu lombu*

LOPAH
SUMATRA
A type of dagger.
(ROGERS)

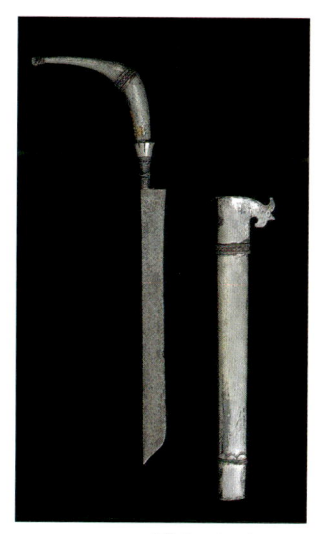

LOPAH PETAWARAN
[GURU-KNIFE, TOMBOLADA, TORDJONG]

SUMATRA, ACEH, GAYO, KARO BATAK.
A dagger with a straight, parallel back and edge. The edge curves at the point towards the back. The hilt is curved at an angle of 90° and has a long thin protrusion. The hilt's form is called *hulu jongo* (meaning: 'shaped as a *jongo*', a species of stork).
The *lopah petawaran* is an ancient type of dagger from which the *rencong* is allegedly derived. It is used, for instance, to make incisions in the ears of buffaloes. After placing this magical weapon in a mixture of water and flour, this liquid may be sprinkled on people or objects in order to neutralise evil influences. *Tawar* means: to make feeble, ward off.
(HEIN 1899; KREEMER; MÜLLER; ROGERS; SIBETH; VOLZ 1912)

322. **Lopah petawaran**
Sumatra, Batak. RMV 3155-20. Acq. in 1954 from W.G. Broek. L. 30 cm.

323. **Lopah petawaran**
Sumatra, Karo Batak. Hilt: *hulu jongo*. L. 29 cm.

324. **Lopah petawaran**
Sumatra, Gayo. Hilt: *hulu jongo*. L. 31 cm.

LOPU
MALUKU, SERAM
A machete with a blade longer than most Indonesian machetes. Its hilt usually has a protrusion at the bottom, to improve its grip. The blade is sometimes indented near the hilt serving to catch the opponent's sword or to fend it off. The *lopu* is rarely carried in a scabbard.
(DRAEGER)

LUDJO ALANG
See *luju alang*

LUDJU ALANG
See *luju alang*

LUDJU NARU
See *sikin panjang*

LUJU ALANG
[LOEDJOE ALANG, LUDJO ALANG, LUDJU ALANG, SIKIN ALANG]

SUMATRA, ACEH, GAYO
A short version of the *sikin panjang*, used as a machete. The term *alang* means 'insufficient (for warfare)'. This sword's edge and back are straight and parallel. The sharp side curves at the point towards the back. Its hilt is always forked or in the shape of an animal's mouth.
(KREEMER; ROGERS)

LUJU ALAS
[LOEDJOE ALAS]

SUMATRA, ACEH, GAYO, ALAS
A short, straight-backed sword with a slightly curved edge. The centre of gravity lies near the point, the edge curves towards the back at the tip. The blade is slightly indented near the hilt called *hulu iku mie*, meaning: 'as the (knotted) tails of cats'.
(KREEMER)

LUJU CELIKO
[LOEDJOE TJELIKO, OELANG ALING, TJOERIGA]

SUMATRA, GAYO
A dagger with a narrow tapering blade. Its hilt is curved to an angle of 90° ending in a short, sharp protrusion. The hilt is made of gold or *suasa*, filled with resin. The scabbard has a small, broad upper part at the opening. The bridegroom carries this weapon on his back in a belt during the marriage ceremony.
(KREEMER)

LUJU LAPAN SAGI
See *sikin lapan sagu*

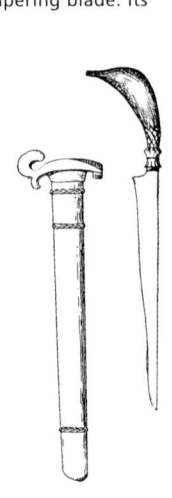

325. **Luju alang**
Sumatra, Tawar Lake district. Hilt: *hulu lunkee rusa*.

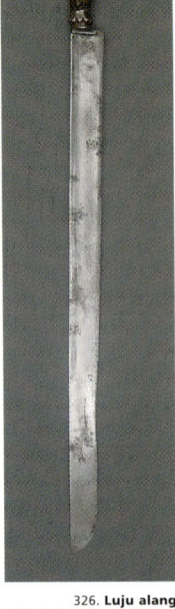

326. **Luju alang**
North Sumatra. Hilt: *hulu rumpung*. L. 61 cm.

327. **Luju celiko**
Sumatra, Gayo. Hilt: *hulu paroh blesekan*.

328. **Luju alas**
North Sumatra. Hilt: *hulu iku mie*.

LUJU MUGENTA
[LOEDJOE MOEGENTA]

SUMATRA, ACEH, GAYO, ALAS
An old-fashioned type of *klewang* with a small bell attached to the bottom of the hilt. Such a sword was carried by front-rank men to inspire their comrades with its sound in combat.
(KREEMER)

LUKI
JAVA
A machete with a fanciful blade and a long, straight hilt.
(RAFFLES 1817)

LUNDJU
See *lunju*

LUNGAT
See *piso raout*

LUNJU
[LUNDJU]

KALIMANTAN
A collective term for spears.
(HEIN 1890)

LUNGKEE
[LOENGKEE, TANDOE, TANDU]

SUMATRA
A type of horn used for manufacturing hilts. Known as *lungkee* (Aceh) and *tandu* (Gayo).
(KREEMER)

LURIS PEDANG
SUMATRA, ACEH
A sword with a narrow blade measuring c.37-75 cm. The hilt has a cross-piece.
(DRAEGER)

LUTONG
KALIMANTAN, KAYAN
A helmet made of plaited rattan covered with scales of a pangolin.
(ROTH 1896B)

LUWUK
[LOEWOEK]

JAVA
A rather short sword with a broad blade. Its edge and back are parallel. The back curves towards the edge at the tip. The hilt is curved and broadens with an indentation at the end.
(JASPER AND PIRNGADIE 1930)

329. **Luki**
Java.

330. **Luwuk**
Java.

331. Keris hilt
Madura. Ivory, decorated with a European crown and winged horses.

332. Mandau
Kalimantan. RMV 5395-16. Acq. in 1985 from Barbara Harrisson. L. 98.5 cm.

MA DADATOKO
See *salawaku*

MAEN
MALUKU, BURU
A type of fighting stick. Shafts of spears may also be used as a *maen*.
(DRAEGER)

MALAB
See *mandau*

MALAT
See *mandau*

MANDAU
[BAIENG, DUKU, KAMPING, MALAB, MALAT, MANDO, PARANG IHLANG, PARANG ILANG]

KALIMANTAN
A sword used when hunting heads, and as a machete in daily life. Characteristic for the *mandau* is that the blade is shaped convexly on one side and somewhat concavely on the other side. Due to this form one finds allegedly two effective directions of striking: (1) in an angle of 45° from the top right hand side to left under side and (2) in the same angle from the left under side to the top right hand side. Although this is a limitation in usage, the striking action would allegedly have much more effect than if the blade were to be shaped identically on both sides. The concave side is where the thumb must be placed. For left handed people the blade would thus be the mirror image of that for right handed people. Its edge is somewhat convexly curved, the back usually straight or somewhat concavely curved. At the tip it curves in an undulating or notched line towards the edge. This part sometimes has small, elegantly forged curls or inlaid brass ornamentation. The blade is, furthermore, often decorated with engravings, encrusted motifs or small circles of copper, brass or silver. These decorations are found, except for the small circles which can be inserted in holes drilled right through the blade, on the convexly curved side of this blade.

The following motifs are found, apart from the small circles:
(a) *mata joh*: S-shaped figures in interlocking spirals and applied to a rather large part of the blade along the back;
(b) *mata kalong*: four mirrored large S-shapes near the hilt;
(c) *tap set sien*: an eight pointed star with a dot in the centre, and to be carried by royals only.

Originally the required amount of iron was melted by the indigenous peoples themselves, later usually European or Chinese iron was used. Blades were, before being used, hardened by heating and then immersing in cold water or by placing them on a red hot iron bar until they showed the correct colouring. The result is a tough and hardened blade.

Strikingly enough, the iron of a fine *mandau* does not oxidise. Sometimes the blade's edge is coloured blue by cutting into the trunk of a capok tree for several hours, and pulling the blade through this incision after each cut. This causes a very beautiful blue colour which remains visible for some six months.

The hilt is usually made of wood (often roots of the *ensurai* tree, *Diptocarpus oblogifolius*) or horn of deer (from the *rusa*, or *sambur*-deer). The hilt has no hand protection. A large section protrudes at the top, towards the edge's side. The protrusion and the hilt's upper part are usually profusely decorated with carvings of *aso*-figures, fanciful faces or other motifs. It is also decorated with tufts of hair. The hilt's central part is wound around with plaited rattan or metal thread. The lowest part is sometimes made of horn and often finished with a thick ring of sticky resin (*damar*) moulded around the base. In this ring a silver coin is placed as a decoration.

333. Mandau
Kalimantan. L. 61 cm.

334. Mandau
Kalimantan. Rear view of illus. 333.

335. Mandau
Kalimantan. L. 69 cm.

336. Mandau
Kalimantan. L. 69 cm.

337. Mandau
Kalimantan. L. 65 cm.

338. Mandau
Kalimantan. L. 68.5 cm.

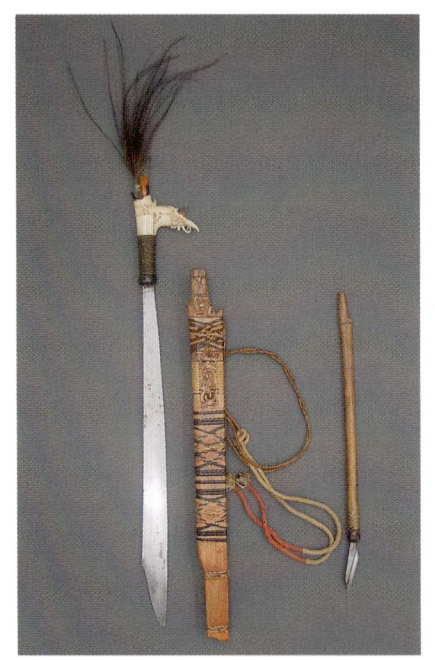

339. Mandau
Kalimantan. L. 64 cm.

340. Mandau
Kalimantan, Kayan. L. 63 cm.

341. Mandau
Kalimantan. Secondary scabbard and knife loose from the scabbard. L. 60 cm.

342. Mandau
Kalimantan. L. 55 cm.

343. Mandau
Kalimantan. L. 69.5 cm.

344. Mandau
Kalimantan. Part of the blade with *mata kalong* and *mata joh* figures.

The scabbard is made of two thin pieces of wood, held together by several broad bindings of plaited rattan to form a knot (*katong evok*). The scabbard is often finely carved and can be decorated by means of, for instance, bone inlays, tufts of coloured hair, beadwork and feathers of the hornbill or argus-pheasant. On the scabbard's back, we see a second, smaller scabbard (*apis*) made of palm leaf or fabric in which a small secondary knife (*piso raout*) is kept. It has a long straight hilt and a short blade, standing at an angle to the hilt and usually like the *mandau* itself, has a convex and a concave side. According to a number of sources this small knife was used, among other things, to remove the soft parts of the heads cut off the enemies.

The *mandau* is worn almost horizontally by means of a rattan girdle (*blavit*) to which all kinds of amulets may be attached. The blade's edge faces upwards. The girdle ends in a noose on one side. On the other side one may find a large 'knot' made of, for example, mother-of-pearl, shell, wood, deer horn, animal teeth, *anggang gading* or some other material, which can be held by the noose. The *mandau* is used by many peoples of Kalimantan and is widely spread.

(AVÉ; CHIN; COPPENS; FURNESS; GARDNER 1936; GOMES; HEIN 1890, 1899; HOSE; LOW; MJÖBERG; RAFFLES 1817; SCHMELTZ 1893; SELLATO; SHELFORD; STONE; TROMP; VOSKUIL 1921-'22)

MANDAU LANGGI TINGGAN
See *langgai tinggang*

MANDAU PASIR
KALIMANTAN
A *mandau* with a very heavy blade.
(STONE)

MANDO
See *mandau*

MARANGI
A mixture of lemon juice and arsenic, used to etch *pamor* weapons.
(BEIDATSCH)

MATANA KNIFE
SULAWESI, MATANA
A knife used in the village of Matana on Sulawesi. Its blade has various shapes and grooves lengthwise. The horn or wooden hilt broadens towards the tip and is curved near the blade.
(GRUBAUER)

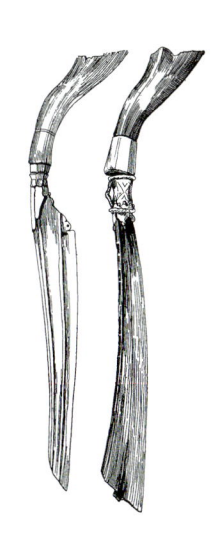

MENTAWA
JAVA
A sabre with a blade of which both the edge and, to a lesser degree, the back are S-shaped. At c.1/3 of length from the tip the back of the blade dips down and then runs almost straight to the tip.
(JASPER AND PIRNGADIE 1930)

MENTOK
JAVA
A short, sturdy sword. The blade's back is slightly curved, the edge somewhat S-shaped. The centre of gravity lies near the tip.
(RAFFLES 1817A; STONE)

MEREMOE
See *amanremu*

MERHOEM
See *bawar*

MERHUM
See *bawar*

MERMO
See *amanremu*

MERMOE
See *amanremu*

MERMOE PAKPAK
See *ladingin*

MERMU
A Sumatran weapon.
(ROGERS)

345. **Matana knives** Sulawesi, Matana.

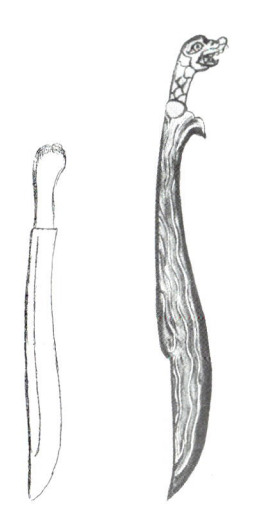

346. **Mentok** Java.

347. **Mentawa** Java.

348. **Mundo** North Sumatra.

MERMU PAKPAK
See *ladingin*

MOENDO
See *mundo*

349. **Moso** Alor. Warrior carrying a sword (*moso*).

MOSO
SULAWESI, ALOR
A sword with a straight blade. The back curves towards the edge at the tip. A small barbed hook is located at the blade's top, where the back begins to curve towards the edge. The hilt has a large triangular end piece, flattened at both sides and decorated with carvings. The scabbard has a broad, rectangular flat mouth-piece and a broadened decorated end.
(FOY)

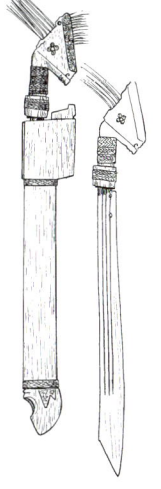

MUNBAT
See *piso raout*

MUNDO
[MOENDO, PARANG SADEUEB]
SUMATRA, ACEH, GAYO, ALAS
Mundo is the term referring to a machete found in Gayo and Aceh which strongly resembles the Javanese *kudi*. Its fanciful blade ends in two strongly curved tips, resembling a bird's beak. In Alas, the *mundo* is the weapon of brides, and also used by women to cut rushes, bushes and thin branches. Its blade is curved, the edge is on the inside.
(KREEMER)

350. **Moso**

NAIBOR
See *niabor*

NARUMO
See *sikin panjang*

NGGILING
See *agang*

351. **Niabor hilt**
Kalimantan, Batang-lupar, Dayak.

352. **Niabor**
Kalimantan. L. 87 cm.

353. **Niabor**
Kalimantan, Batang-lupar, Dayak.

NIABOR
[BEADAH, NAIBOR, NYABOR, NYABUR, PARANG NJABUR LAKI-LAKI]

KALIMANTAN, SEA DAYAK, IBAN

The *niabor* is a curved sword. Its blade has a convex edge and a concave back broadening towards the tip so that the centre of gravity lies at the point. The edge curves in a faint curve towards the tip. The blade is usually c.60 cm long, sometimes even 90 cm. On the blade's sharp side, near the hilt, we find a blunt part with usually a protrusion (*kundieng*) in the middle, serving as finger protector. This protrusion is typical of the *niabor*. The blade is rectangular under the finger protector. This part of the blade is called *sangau*. The part of the blade which is called *tamporian* is located between the finger protector and the hilt. This part is either rounded or poly-angular in cross-section. The blade is rarely decorated. Sometimes, however, a rib runs along the rim, on both sides of the blade from the *tamporian* until close to the tip.

The hilt is made of deer horn (of the *rusa* or the *sambur* deer) or of wood. Its strongly flattened top is decorated with carvings which may have the following patterns:
- (a) *cantok resam* (shoots of the *Gleichenia dichtoma*);
- (b) *telingai* (meaning: 'scorpion');
- (c) *entadok kaul* (meaning: 'interlocking caterpillars').

If the hilt is made of deer horn, the same part of the antlers is used as with the *mandau*. The blade is attached to the cut off end of the main antler branch. Hilts may have metal rings (*grunieng*) and end in a long protrusion which curves down towards the blade's sharp edge. The hilt and scabbard are never decorated with hair.

(GARDNER 1936; HEIN 1899; SELLATO; SHELFORD)

NIO
NIAS

On Nias *nio* is the general term used for a hilt. The most frequently used material to make hilts is wood, with a metal cylinder around the lowest part. Besides this, brass or bronze is used to produce the hilts of smaller swords. Hilts of bone or buffalo horn, imported elephant tusk or molar-teeth are also found, but are rather rare. See also *bekhu* and *lasara*.

The following types of *nio* occur:
- (a) the *niobawa lawolo* is cut in the form of a *lasara's* head whose open mouth, with teeth, is the most striking element, with the *bekhu* figure lying on the neck of the *lasara*. The *niobawa lawolo* is most frequently found on Nias and comes in many variants, from realistic to rather abstract;
- (b) the *nioasu buwuna*, the name on south Nias for a *lasara* variant. The *bekhu* figure is not found on this type of hilt;
- (c) the *niobawa bae* representing the head of a monkey often with the bekhu on its neck. Due to the locked teeth the *niobawa bae* appears very aggressive. The monkey allegedly provides the weapon with speed in combat;
- (d) *nioio gari* is shaped as a very stylised *lasara*. It ends in a wide-open, V-shaped mouth. The *lasara's* eye is placed against the corners of the mouth. From this mouth comes a long, thin small iron stave which is curved at the tip. The rather rare swords with this type of hilt are called *gari* and are reserved for a headman as they indicate a high status;
- (e) the *niofo m'bowaja* is found in north Nias;
- (f) the *niofabawa lawolo* is found in south Nias and is made of brass or bronze only. It has the shape of a very stylised animal's head, ending in a V-shaped mouth with several sharp points representing teeth;
- (g) the *niotaka waena*, the south Nias term for 'a hilt as a swallow'. The lips of the *lasara* mouth are stylised, flatly cut, and recede completely;
- (h) the *nioto lutolu* (meaning: 'a hilt in the form of a beetle') resembles (g) but comes from north Nias. The flat, retreating sides are allegedly borrowed from the spread wings of this flying insect and serve to give speed to the sword;

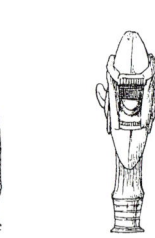

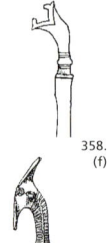

354. (a)
355. (b)
356. (c)
357. (d)
358. (f)
359. (g)
360. (h)

(i) the *niokawa kawa* (meaning: 'a hilt in the form of a butterfly') resembles (g) and (h). It represents a moth (a nocturnal insect), making the weapon inaudible during nightly raids;

(j) the *niobu kaka* has the shape of a bird's head. It is found only in brass or bronze and resembles Javanese *keris* hilts;

(k) the *niowoli woli* has the shape of an open fern leaf. It is mainly found on smaller knives which are usually used for slicing *pinang* nuts;

(l) the *nioloa uma* is also known as *nio bawa manu*. These two names are indeed given to one and the same type of hilt described as 'in the form of a convex hammer'. Such hilts almost only occur with imported elephant tusks, often coloured reddish. They are found on the *si euli*, the *saboa io* of northern Nias and the *balatu nifoio* of southern Nias. These daggers seem to be a combination of the *rencong* and the *sewar* of Aceh;

(m) the *niodanga wana* is a very simple, rigid and stylised wooden hilt in the shape of a butt-end of a gun. It is often found on the *saboa io*, the *balutu nifoio* and on smaller working knives.
(TEXT: K.H. SIRAG)

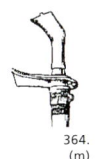

361. (j)
362. (k)
363. (l)
364. (m)

NUME
FLORES
A machete used in combat. It resembles a halberd with a long, broad, slightly curved blade and a long hilt or shaft.
(DRAEGER)

NYABOR
See *niabor*

NYABUR
See *niabor*

O GOHOMANGA MA URU
MALUKU, HALMAHERA, TOBELO
The wooden hilt of the *o humaranga*, a sword used by the Tobelo on Halmahera. The hilt curves halfway in an angle of c.45°. It may be inlaid with pieces of white earthenware and mother-of-pearl of the *nautilus* shell. Its end is cut-out to represent an open crocodile mouth.
(PLATENKAMP)

O HUMARANGA
[TURI]
MALUKU, HALMAHERA, TOBELO
The foreign origin of these swords is expressed in the hilts in which material from abroad such as mother-of-pearl of the *nautilus* shell or pieces of broken china are used. These hilts may depict a 'bird's beak' or a 'crocodile mouth' (*o gohamanga ma uru*), have deeply cut out ends, and are often carved by the Tobelo themselves. These motifs may represent animal ancestors.
In the past, these (blades of) the swords, which broaden towards the point, were either part of war spoils or purchased from non-Tobelo traders. Even nowadays (1990) the Tobelo do not forge the blades themselves. This is done by smiths from elsewhere who have settled down in the region in the course of time. The form is, however, drawn by the Tobelo taking an existing sword as an example. The sword is named after its blade. Apart from the *o humaranga*, we see the 'flower of the *turi*' and the '*rinomu*'. The *o humaranga* is the most prestigious type of weapon. To forge its blade, a plate spring of a car is sometimes used. The sword's measurements are set against those of the human body. The blade's length may not be precisely the same as the sum of a number of hand palms. Its width may not be equal to the length of an index finger. The hilt is carved out of wood and blackened with a mixture of soot and plant juice. It may be decorated with carvings of simple cross and hour-glass motifs. Hilts inlaid with white earthenware are also found. The blade's back runs in an oblique angle towards the point.
The name *o humaranga* points at the town Semarang on Java, from which these swords allegedly originate. They are extremely suitable as a marriage gift: the bride-taker should present one to the bride-giver. They are worn during the *hoyla* dance, the battle-dance included in the Tobelo marriage ceremony.
(PLATENKAMP)

OEDJONG TIPIS
See *rudus*

OELANG ALING
See *luju celiko*

OELEE BABAH BOEJA
See *hulu babah buya*

OELEE BOH GLIMA
See *hulu boh glima*

OELEE DANDAN
See *hulu dandan*

OELEE DJONGO
See *hulu jongo*

OELEE GEUREUPONG
See *hulu puntung*

OELEE IKOE ITE
See *hulu iku ite*

OELEE IKOE MIE
See *hulu iku mie*

OELEE LAPAN SAGOE
See *hulu lapan sugu*

OELEE LOENGKEE ROESA
See *hulu lungkee rusa*

OELEE MEU APET
See *hulu meu apet*

OELEE MEUTJANGGE
See *hulu meucangge*

OELEE PAROH BLESEKAN
See *hulu paroh blesekan*

OELEE PEUDADA
See *hulu peudada*

OELEE PEUSANGAN
See *hulu peusangan*

OELEE POENTONG
See *hulu puntung*

OELEE ROEMPOENG
see *hulu rumpung*

OELEE TAPA GOEDA
See *hulu tapa guda*

OELEE TJANGGE GLIWANG
See *hulu cangge gliwang*

OELEE TOEMPANG BEUNTEUENG
See *hulu tumpang beunteueng*

OELOE DENDON
See *hulu dandan*

OELOE GEREPOENG
See *hulu puntung*

OELOE GLIMO
See *hulu boh glima*

OELOE LAPAN SAGI
See *hulu lapan sagu*

OELOE MOEAPIT
See *hulu meu apet*

OELOE OEKI N ITI
See *hulu iku ite*

OELOE POENTOENG
See *hulu puntung*

OELOE SERAMPANG
See *hulu babah buya*

OELOE SIMPOEL
See *hulu iku mie*

OELOE TANDOE N AKANG
See *hulu lungkee rusa*

OELOE TAPA KOEDO
See *hulu tapa guda*

OELOE TJOLO
See *hulu cangge gliwang*

OELTOEP
See *ultup*

OLA
See *awola*

OPI
WETAR AND KISAR ISLANDS
A sword the blade of which has an almost straight edge and back broadening towards the tip. The back runs in an oblique angle towards the edge. The hilt is usually made of horn. Its large knob is decorated with long strands of hair. The scabbard, *opi kapan*, has a broader mouth to one side and a somewhat protruding, elegantly carved end.
(FOY; STONE)

365. **Opi**
Wetar. L. 57.5 cm.

366. **Opi**
Alor (?). L. 52.5 cm.

367. Klewang hilt
Timor. RMV 360-5584. Acq. in 1883 through the Koninklijk Kabinet van
Zeldzaamheden. 's-Gravenhage. L. 113 cm (entire weapon).

368. **Pedang lurus**
Java. RMV 1848-3. Acq. in 1913 from dr. A.W. Pulle. L. 56.5 cm.

P
p
ALPHABETICAL SURVEY

PACRET
[CECRET, PATJRET, TJETJERET]
SUMATRA, ACEH
A bamboo squirt to spray a mixture of water and pepper at the enemy to render them helpless. This device was also used by thieves. The Gayo term is *cecret*.
(KREEMER)

PADANG SABIT
SUMATRA, MINANGKABAU
A sickle with a long hilt with a curved end standing at an angle of almost 90° to the blade, of which the edge is straight and the back slightly curved.
(DRAEGER)

PADE
See *labo*

PADIMPAH
SOUTH-WEST SULAWESI, TORAJA
A combat weapon made of hardwood, resembling a boomerang, slightly curved whereby the centre is flat and both ends round in cross-section. The method of use is unknown.
(DRAEGER)

PAKAYUN
KALIMANTAN, MURUT
A sabre characteristic of the Murut, a people inhabiting parts of north Kalimantan. It has a long narrow blade (c.60-65 cm long and 3 cm wide) without decoration. It is slightly curved and of the same width over the entire length.
The hilt is always made of wood and ends in two parallel points, standing obliquely on the hilt. The space between these points may have carvings filling up a part of the space in between or in some cases even reaching past the points. The hilt has a brass cylinder widening at the blade to a round rim (*umbo*) which serves as a finger protector. This cylinder seldom reaches the fork. Plaited rattan covers the space between the cylinder and the fork.
The scabbard is usually coloured red and white. It is made of two wooden boards held together with rattan, thread or tin plates. The spaces between these bindings have geometrical figures on the outside. On the inside a small pouch of tree bark, decorated with hair, may be found.
(GARDNER 1936; SHELFORD; STONE)

369. **Pakayun**
Kalimantan, Murut. L. 81.5 cm.

PALANGGE
SUMATRA
A wooden sword used to exorcise spirits.
(ROGERS)

PALANGKA
See *sikin pasangan*

370. **Palitai**
Mentawai, Siberut. L. 33.5 cm.

PALITAI
[PALITTEI, PARITTEI, PATTEI]
MENTAWAI ISLANDS
A knife with a smooth blade both edges of which are sharpened and run parallel. The edges come together at the tip to end in a sharp point. The blade has in the middle along the entire length an elevated rib. The steel used to produce blades was imported from Sumatra, as forging was unknown on Mentawai. The blades were finished in the desired form on the spot. The total length may vary from c.30 cm to 1 m.

The hilt of the *palitai* is thin and long, at the blade still rather broad but becoming thinner to run long and elegantly into an almost sharp tip or a decorated end. This hilt has a round thicker part just past half way. On the northern islands (Siberut, Sipora) the hilt is rather simple, on the southern islands (north and south Pagai) it is curved more elegantly in the shape of a swan's neck with a curl (similar to a closed fern leaf) at the end. Other decorations, such as a stylised head of a cock or bird are found, too. This hilt may have carvings filled with resin. The scabbard has the same shape as the blade and may be smoothly finished. However, decorations with chicken feathers and rattan ties at the mouth also occur. Sometimes the scabbard has a bone or deer-horn tip.

The *palitai* is carried on the right, in the loin-cloth and may be part of the dowry.
(FABER; SCHEFOLD; STONE; VOSKUIL 1921; WIGGERS)

371. **Palitai**
Mentawai. RMV 3600-1209. Acq. in 1959 through the Ethnografisch Museum van de Koninklijke Militaire Academie, Breda. L. 42.5 cm.

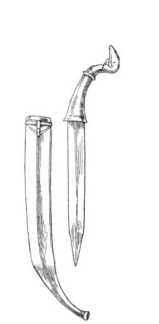

372. **Palitai**
Mentawai, Siberut, Katorei. RMV 79-8. L. 34.5 cm.

373. **Palitai**
Mentawai. RMV 835-19. L. 32.5 cm.

PALITTEI
See *palitai*

PALUON
KALIMANTAN, MURUT
A quiver used for transporting blow-darts. Both the quiver and the lid are made of bamboo, strengthened by means of plaited rattan strips. The quiver has a long wooden hook with which to attach it to a girdle. The top of this hook may be decorated with a carved motif.
(SELLATO)

PAMANDAP
SUMATRA, MINANGKABAU
A sword carried at the side of the body.
(DRAEGER; MARSDEN)

PAMOELOE
See *pamulu*

PAMOR
A term referring to a pattern in the blade made by means of forging together various metals. By repeatedly forging, folding double, twisting, cutting into bits and re-forging etc., pre-determined patterns appear in the metal. These are made visible through the effect of acid, usually combined with a compound of arsenic. The most beautiful *pamor* is made by using normal iron in combination with iron containing nickel. Once coloured with acid containing arsenic, the normal iron turns black while the nickel-holding iron remains silver-coloured. The pattern in the metal is thus beautifully brought to the fore. Nickel-containing iron originates from meteorites. Alloys used later contain nickel from other sources such as bicycle scrap-iron. Sometimes even pure nickel is used. Besides, instead of nickel-containing iron combined with (non-)nickel-containing iron, iron of various quality and hardness was used. After applying acid, the colour does not change, but the *pamor* pattern appears in relief. Many Javanese *pamor* motifs are found, each with its own meaning. The majority is derived from, or based on, the:

(a) *pamor wos wutak*, meaning: 'scattered rice grain';
(b) *pamor sekar pala*, meaning: 'nutmeg blossom';
(c) *pamor sekar ngadej*, meaning: 'blossom standing upright';
(d) *pamor blarak ngirid*, meaning: 'parallel coconut leaves';
(e) *pamor sekar temu*, meaning: 'ginger blossom'.

(VAN DUUREN 1996A, 1998; FREY; GARDNER 1936; GRONEMAN 1908, 1904A-E, 1910A; HARSRINUKSMO; HILL; HUYSER; JASPER AND PIRNGADIE 1930; JENSEN; KERNER; SOLYOM; TAMMENS)

PAMULU
[PAMOELOE]
SULAWESI
A spear with a very thin, narrow iron or copper blade in the shape of a nail. The blade may have some undulations at the tip. The shaft may have a *baranga* (Buginese) or a *banrangang* (Makassar) i.e., an attached decoration near the base consisting of horse's or goat's hair, or the feathers of a cock.
(MATTHES 1874, 1885; SCHRÖDER)

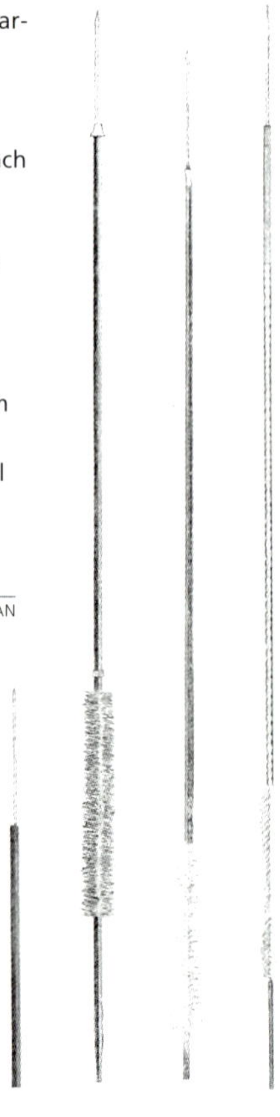

374. **Pamulu** Sulawesi.

PA-NA
See *li-pun*

PANA
[PANAH]
JAVA
An arrow used by Indo-Javanese deities, demigods and heroes.
(RAFFLES 1817A)

PANAH
See *pana*

PANAH AYER
A squirt containing a poisonous juice of the *buta buta* tree (*buta*, meaning: 'blind') which can blind the enemy.
(GARDNER 1936)

PANDAK
See *katok II*

PANDAT
[PARANG PANDIT]
KALIMANTAN, SARAWAK, LAND DAYAK, BIDAYUH
The *pandat* is the war sword of the Land Dayak of Sarawak in northwest Kalimantan and is never used as a tool. It is handled with two hands, with a downwards stroke. Its blade and hilt are forged from one piece. The blade is bent just below the hilt at an angle of c.25°. The bend in the blade is located in the transitional part between the blade and the hilt. Both the back and the edge are straight and run apart, so that the blade's broadest part is at the point.

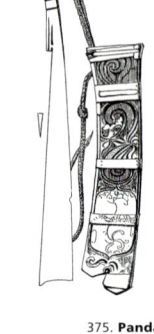

375. **Pandat** Kalimantan.

376. **Pandat** Kalimantan, Kapuas River.

This point of the *pandat* may have the following shapes, whereby:
(a) the blade's back is longer than its edge. In the oblique plane thus brought into being a V-shaped indentation is made at the tip, closer to the edge than to the back. The shortest side of the V is rounded, the longest side may be somewhat curved. This point is characteristic of the Sidin Land Dayak people;
(b) the blade's back is shorter than the edge. The above-mentioned V-shaped indentation lies closer to the back than to the edge;
(c) the ends of the inner angle are formed into short hooks or protrusions in which brass nails (*lantak paku*) have been driven into pre-fabricated holes in the blade, near the tip. This form is characteristic of the Bennah Land Dayak;
(d) the blade's back is of the same length as the edge. The sides of the above-mentioned the V-shaped indentation are of equal length.

The blade's cross-section varies along the entire length. Near to the hilt the form is rectangular, a flat wedge shape at the point. The hilt is rectangular in cross-section along the entire length and is forged a little narrower just above the middle. The hilt's width is the same along the entire length. Just above its centre a hole is drilled through which a short iron bar (*sekak*) is placed. At the end of the hilt a sharp point is forged covered by a decoration of horn or brass. The hilt may be decorated with brass strips or with silver or tin foil. On the back, a tuft of hair may serve as a decoration. Near the hilt some grooves may be made in the blade. The hilt's upper part is

377. Pandat. Kalimantan. L. 56 cm.
378. Pandat. Kalimantan. L. 68.5 cm. 
379. Pandat Kalimantan. L. 72.5 cm. 
380. Pandat. Kalimantan. L. 59 cm. 

held in one hand, with the index finger over the *sekak*, gripping the lower part, held with the other hand.
The scabbard is made of two parts held together by plaited rattan strips or by metal strips. Both the back and the edge may be finished with bamboo strips. The scabbard is equally broad over nearly the entire length, only the mouth is somewhat broader. This scabbard may be decorated with incisions or carvings in geometrical, plant or animal motifs. These figures may be filled with tin foil and may be applied to one or both sides. The scabbard is sometimes decorated with small tufts of hair or feathers and, furthermore, may be painted red.
(CHIN; GARDNER 1936; LOW; ROTH 1896B; SHELFORD; STONE; VAN ZONNEVELD 1988)

PANDJI
See *panji*

PANGHO
SULAWESI, TOMORI
The sword of the Tomori people. Its *pamor* is made of a lightly coloured kind of iron found in the mountains of Torongku and Ussu in the Luwu realm along Boni Bay. The iron mixed with this lightly coloured iron may be found in the mountains around Matano Lake.
(GRONEMAN 1904C)

PANGOER
See *pangur*

PANGUR
[PANGOER]
SUMATRA, ALAS REGION
A dagger, usually with a forked hilt.
(VOSKUIL, ROGERS)

PANGUR CUT CUT
[PANGUR TJUT TJUT]
SUMATRA
A type of knife.
(ROGERS)

PANGUR TJUT TJUT
See *pangur cut cut*

PANJI I
[PANDJI]
SUMATRA, KARO BATAK
A war banner of white cotton which is attached to a long pole consisting of two pieces. The lower part is made of black wood, the upper part is of bamboo. The *panji* is covered with motifs and letters. The point is decorated with feathers of a cock.
(VOLZ 1909)

PANJI II
See *ranjau*

PANJIE
See *ranjau*

PAPATN
See *utap*

PARABAS
KALIMANTAN, DAYAK
A type of trap. A tree is almost felled and kept upright with rattan ropes. When the enemy approaches the ropes are cut so that the tree strikes the enemy.
(STONE)

PARANG
[BERANG]
A collective term for swords and machetes hailing from all over the archipelago.

PARANG BAKONG
See *sadeueb*

PARANG BANKONG
A type of *parang*.
(GARDNER 1936)

PARANG BEDAK
KALIMANTAN
A short sword with a heavy blade sharpened on one side. The edge is convex, the back is straight until the tip where it curves.
(STONE)

381. Panji Sumatra, Karo Batak.

PARANG BENGKOK
JAVA, BALI
A heavy machete with a back between straight and convex, curving strongly towards the tip. The edge is between S-shaped and concave. The blade's tip is curved sharply upwards.
(GARDNER 1936)

PARANG CANDONG
[PARANG TJANDONG]
SUMATRA
A machete with a curved point.
(ROGERS)

PARANG CHAKOK
A variant of the *parang*, mainly used for pruning trees.
(GARDNER 1936)

PARANG CHANDONG
A variant of the *parang*.

PARANG CHERIGA
See *parang rantai*

PARANG DJIMPUL
See *jimpul*

PARANG GABUS
A variant of the *parang* made of strong, hardened steel.
(GARDNER 1936)

PARANG GEDAH
SUMATRA, ACEH
A curved sword the blade of which has a clearly concave back and a convex edge. This blade broadens to the tip and ends in a semicircular point. The hilt is remarkably long.
(NIEUWENHUIZEN)

PARANG GEUDANG
[GEUDANG]
SUMATRA
A type of machete.
(ROGERS)

PARANG GINAH
A short, concavely curved sword, presumably used as a sickle. The edge is concave, the back convex with about halfway along a small upright tooth. The hilt is long and straight.
(STONE)

PARANG GONDOK
A variant of the *parang*.

PARANG IHLANG
See *mandau*

PARANG IKOE LINONG
See *parang iku linong*

PARANG IKU LINONG
[IKOE LINONG, IKU LINONG, PARANG IKOE LINONG]
SUMATRA
A type of machete.
(ROGERS)

382. **Parang bengkok**

PARANG ILANG
See *mandau*

PARANG KAJOELIE
See *parang kajuli*

PARANG KAJULIE
[PARANG KAJOELIE]
KALIMANTAN
A knife with a blade resembling that of the *barong*.
(STONE)

PARANG KOTENG
A variant of the *parang*. Its blade and hilt are made from a single piece.
(GARDNER 1936)

PARANG LADING
A variant of the *parang*, the back of which can be straight or concave. The edge can be straight or convex. Near the tip the blade broadens. The tip itself is straight and stands at right angles to the back edge or runs somewhat obliquely. The edge is longer than the blade's back.
(GARDNER 1936)

PARANG LATOK
[LATOK, LATOK BUKU, PARANG PATHI]
KALIMANTAN
A sword, also used as a machete. Its heavy blade shows a bend near the hilt. The part between the hilt and the bend is usually rectangular in cross-section, but polygonal or rounded shapes also occur. The blade broadens towards the point, whereby the back at the tip curves towards the edge. The blade's back is thick causing the blade to be wedge-shaped in cross-section.
The entire hilt is round in cross-section, and broadens towards the top to a knob, flattened on both sides. This knob, usually made of wood, has cross grooves. The hilt's round part is often strengthened with a cord or plaited rattan, sometimes with silver rings or a silver sleeve.

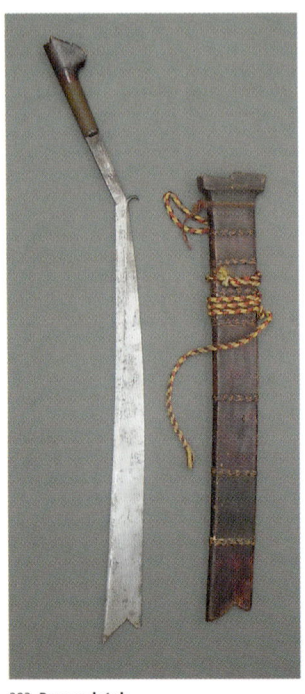

383. **Parang latok**
Kalimantan. L. 70 cm.

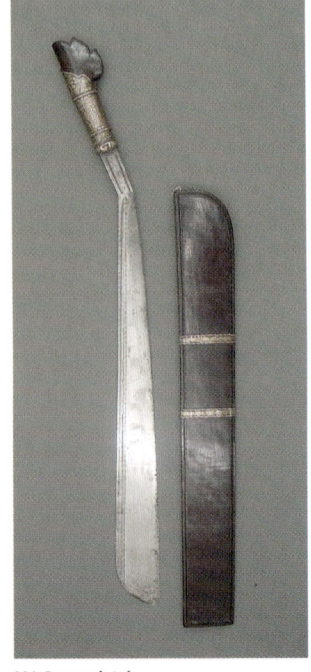

384. **Parang latok**
Kalimantan. L. 66.5 cm.

ALPHABETICAL SURVEY

The scabbard covers only the blade's lowest part, reaching the bend. The *parang* is used two-handedly, whereby one hand holds the hilt and the other the blade's shoulder, in order to strike downwards. The *parang latok* is larger, but for the rest almost the same as the *buko*. The *sadap* is a variant of the *parang latok*. The shoulder of its blade is octagonal in cross-section.
(DIELES; GARDNER; SHELFORD; STONE)

PARANG LEUNTEK
[LEUNTEK]

SUMATRA

A type of machete.
(ROGERS)

PARANG LOTOK
[PARANG PATANI]

A short, heavy machete with a hook-shaped point.
(GARDNER)

PARANG NABUR
[BELABANG, BELADAH]

KALIMANTAN

A sword with a curved blade broadening towards the point. The edge is convex, the back concave. The edge may bend towards the back or the back may bend towards the edge at the point. The hilt is usually made of bone or horn, sometimes of wood, and often has a protection for the hand and fingers made of brass or iron. The scabbard follows the blade's shape.
(STONE)

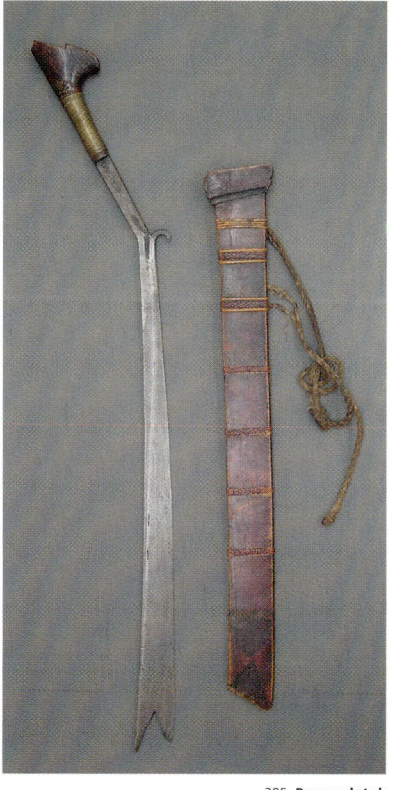

385. **Parang latok**
Kalimantan. L. 81.5 cm.

PARANG NEGARA
A variant of the *mandau* with a ribbed blade.
(STONE)

PARANG NJABUR LAKI-LAKI
See *niabor*

PARANG ONGKOK
A variant of the *parang* with a remarkably concave blade.
(GARDNER 1936)

PARANG PAJAH
A variant of the *parang* with a remarkably concave blade.
(GARDNER 1936)

PARANG PANCONG
A variant of the *parang*.
(DIELES)

PARANG PANDAH
A short, heavy *parang* with a slightly concave blade.
(GARDNER 1936)

PARANG PANDIT
See *pandat*

PARANG PANGGONG
A long, sword-like variant of the *parang*.
(GARDNER 1936)

PARANG PANJANG
SUMATRA, BATAK, ACEH, JAVA

A sword with a straight edge and back, coming together in a rather blunt point. The blade's edge may have a protrusion near the hilt.
(GARDNER 1936)

PARANG PARAMPOEAN
See *parang parampuan*

PARANG PARAMPUAN
[PARANG PARAMPOEAN]

KALIMANTAN

A variant of the *klewang*.
(STONE)

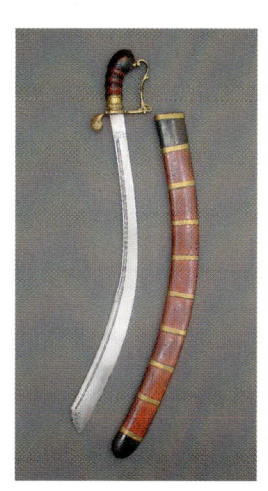

386. **Parang nabur**
Kalimantan. L. 85 cm.

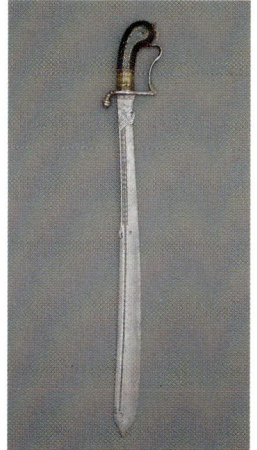

387. **Parang nabur**
Kalimantan. L. 69 cm.

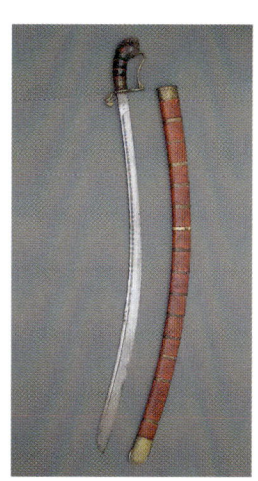

388. **Parang nabur**
Kalimantan. L. 104 cm.

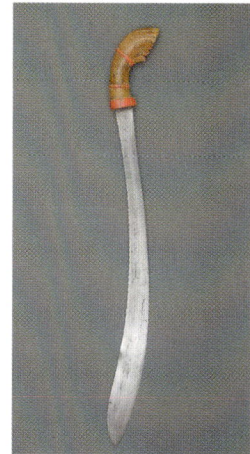

389. **Parang nabur**
Kalimantan. L. 61 cm.

390. **Parang nabur**
Kalimantan. L. 70.5 cm.

PARANG PATAH
KALIMANTAN
A sword used by the Land Dayak of west Kalimantan. At c.1/4 of the length from the hilt, the blade bends back at an oblique angle. From this bend on, several grooves run along the blade. Its remaining part is almost straight, broadening somewhat towards the point. The back curves in an elongated S-shape towards the edge. At the top of the hilt we see a slanting, richly decorated knob.
(AVÉ)

PARANG PATANI
See *parang lotok*

PARANG PATHI
See *parang latok*

PARANG PEDANG
KALIMANTAN, SARAWAK
A sword used by both the Malayan people and the Milanos, a coastal people converted to Islam. The *parang pedang* is mainly used for chopping in the jungle or for splitting the sago-palm. It may also serve as an effective weapon. The blade, measuring c.60 cm, is strongly curved and has no decorations. A very broad part (c.6.5 cm) is found from c.2/3 of the way along the blade from the hilt. The edge continues almost up to the hilt.
The hilt used for the *parang pedang*, the *parang latok* and the *buko* is always made of wood, and has a knob towards the blade's edge, with grooves across it. The knob's sides are flattened. The hilt can be covered with plaited rattan. The scabbard is quite plain.
(GARDNER 1936; SHELFORD)

PARANG PENDAK
A variant of the *parang*.
(DIELES)

PARANG PERANNGI
A variant of the *parang*. Its hilt and blade are made of one piece.
(GARDNER 1936)

PARANG RANTAI
[PARANG CHERIGA, TJOERIGA]
SULAWESI
An ancient type of *parang* with an iron hilt and a heavy, broad blade. At the end of the blade and of the hilt, as well as at the junction between blade and hilt, iron rings are attached to which chains are fastened. The *parang rantai* is said to have magic powers bringing good luck, and can cure the sick when it is placed underneath the bed. The *parang* was also placed on the bodies of the deceased.
(GARDNER 1936)

PARANG SADEUEB
See *mundo*

PARANG SA-KAMPOK
A machete of hardened and very sharp steel, used as a weapon in the jungle.
(GARDNER 1936)

PARANG SARI
A machete-*cum*-weapon with an undulating blade and a partially blunt edge (*sampai*). In ancient stories it is called *ada parang sari sa-belah tujoh-belas lok-nya*, meaning: 'once upon a time there was a *parang sari* with seventeen waves in the blade'.
(GARDNER 1936)

PARANG TEUPAT
[TEUPAT]
SUMATRA
A type of machete.
(ROGERS)

PARANG TJANDONG
See *parang candong*

PARANG UPACARA
[PARANG UPATJARA, UBLAKAS]
SULAWESI
A short machete with a heavy blade and found in various shapes.
(DRAEGER)

PARANG UPATJARA
See *parang upacara*

PARINSE
See *peurise*

PARISE
See *peurise*

PARITSE
See *peurise*

PARITSE DONDANG
See *peurise kajee*

PARITTEI
See *palitai*

PASEKI
SULAWESI, MINAHASA, TOMINI BAY
An ancient type of metal helmet of European origin, often dating from the 17th century and generally following the model of a brass helmet of Spanish design, as used by the VOC (Dutch East Indies Company, 1602-1795).
(DRAEGER; GRUBAUER; VOSKUIL)

PASER
JAVA
A poisonous dart for a blow-pipe (*tulup*). According to Raffles not used on Java for centuries.
(RAFFLES 1817A; STONE)

PASPATI
JAVA
An arrow with a crescent-shaped point used by Indo-Javanese demigods and heroes.
(RAFFLES 1817A; STONE)

PATJRET
See *pacret*.

PATOBANG
KALIMANTAN, DAYAK
A covered pitfall trap, c.1 metre deep, with sharp stakes set in the bottom. Such traps were used to defend forts and villages.
(STONE)

PATREM
JAVA
A dagger with a narrow, pointed blade owned by women. Both sides of the blade are straight and come together in a curve. At

the top the blade has a broadened cross-piece exactly fitting the scabbard. The hilt of the *patrem* resembles a *keris* hilt.
(JESSUP)

PATTEI
See *palitai*

PECUT
[PETJUT]
LOMBOK
A bludgeon measuring *c*.1 m in its entirety. It is made of a long hardwood hilt to which a ball of knotted leather or metal is tied. This ball may be directly attached to the hilt, or to a strip of leather between hilt and ball.
(DRAEGER)

PEDA
NORTH SULAWESI, MALUKU
A sword made in Kaidipang and Buool (north Sulawesi) since the end of the 18th century. It has a rather short blade and a straight back and edge, running apart towards the end. Near the point the back curves to the edge in an obtuse angle. Hilts from Sulawesi are curved at an angle of almost 90° to the blade, and are made of horn. Their scabbards are completely straight. The Moluccan *peda*, however, has a shorter, broader blade and a less curved hilt.
(FOY; HILKHUIJSEN)

391. Pedang I
Sumatra, Aceh. RMV-3600-410. Acq. in 1959 through the Ethnografisch Museum van de Koninklijke Militaire Academie, Breda. Hilt: *hulu meu apet*. L. 97 cm.

PEDANG
[PEDEENG, PEDENG, PEUDEUENG, PODOENG, PODUENG]
The name *pedang* is a collective term used for sabres and swords of various origins in the Gayo and Alas regions of Sumatra. If not described below in separate entries with their specific name (*pedang* followed by an addition), they can be dealt with as follows:

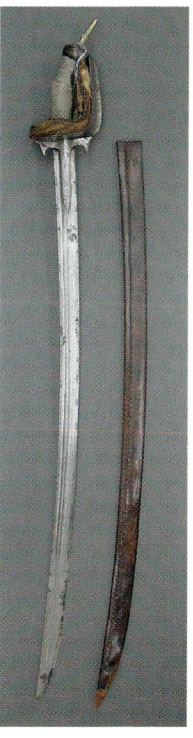

392. Pedang I
Sumatra, Aceh. Hilt: *hulu meu apet*. L. 93 cm.

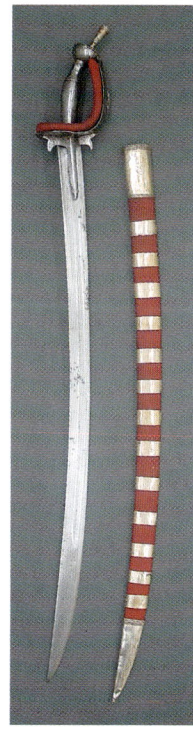

393. Pedang I
Sumatra. Hilt: *hulu meu apet*. L. 88 cm.

394. Pedang I
Sumatra. Hilt: *hulu meu apet*. L. 96 cm.

PEDANG I
SUMATRA
A curved *pedang* of mainly foreign fabrication. The people of Sumatra regard it as less suitable for combat. They prefer the point of gravity to be nearer the blade's point. This *pedang* has for the most part become a ceremonial weapon, and thus an expensive item.

The blade is usually of European, Turkish, Arabic or Anglo-Indian provenance. It may be found in its original form, sometimes the point has been altered. Blades resembling foreign forms were also made in Sumatra, sometimes even imitating the blade's factory-mark. Native blades often have a layered structure and are slightly curved, the back concave, the edge convex. Parts of both sides of the blade are concave, so that parallel to the entire edge and part of the back shallow ribs are formed coming together near the point. Along the blade's back we usually see one two or three grooves differing in length and depth. Some blades have three narrow grooves of different lengths and no concave parts in their sides.

The hilt, *hulu meu apet* (Aceh), has a sharp protrusion at the end. It is usually made of iron and is oval or round in cross-section, with a knob at the top. Sometimes a small golden crown (*tampo*) stands at an angle to the hilt. This pommel has a pointed projection standing at the same sharp angle to the hilt. The term for a *pedang* with a golden knob is *peudeueng meutampoh*. The *hulu meu apet* has a basket-guard covered on the inside with a small pillow (*bantaj*). This guard includes an iron plate with a sharp ridge forged along the middle. Underneath the basket, firmly riveted to it, a metal bar with curled projecting ends is found. From this bar, on both sides of the blade, we see two flattened pointed projections which either grip the scabbard firmly, or fit precisely into indentations made in the top of the scabbard. The hilt is wound around (*teurhat*) with woven metal thread, *kabat* (named *kobot* in the Gayo region) of silver or gold (*teurhat meueh*). From *c*.1900 onwards the threads around the hilt are more densely woven than on earlier examples. At that time the skilled goldsmith T. Nja Buntang significantly improved the quality of the weaving. The *kampong* Garot (Pidie) was then famous for the production of *teurhat*. The scabbard (*sarung*) of Sumatra is made entirely of buffalo hide, or of two wooden strips. It is usually completely covered with red or black coloured buffalo hide or with red cloth. If there is no cover, metal strips hold the halves of the scabbard together. Its upper and lower parts are usually covered with metal, but sometimes decorated with filigree figures, precious stones and embroidery. If the metal parts are made of gold, the weapon is called *peudeueng sarong meueh*. Sometimes the scabbard has a separate mouth-piece (*jambang*) made of wood or ivory.

(DRAEGER; GARDNER 1936; JACOBS; JASPER 1904; JASPER AND PIRNGADIE 1930; KREEMER; KRUYT; VAN LANGEN; MARSDEN; NIEUWENHUIZEN; SNOUCK HURGRONJE 1914; VELTMAN; VERMEIREN 1987; VOLZ 1909, 1912; VOSKUIL 1921)

PEDANG II
SUMATRA

A slightly curved sabre with a hilt in the shape of a *wayang* figure. The scabbard's upper part has a somewhat broader mouth, usually protruding, especially on the blade's edge side.
(FISCHER 1918)

395. **Pedang II**
Sumatra, Palembang. L. 75.5 cm.

396. **Pedang III**
Hilt of illus. 397. L. 14 cm.

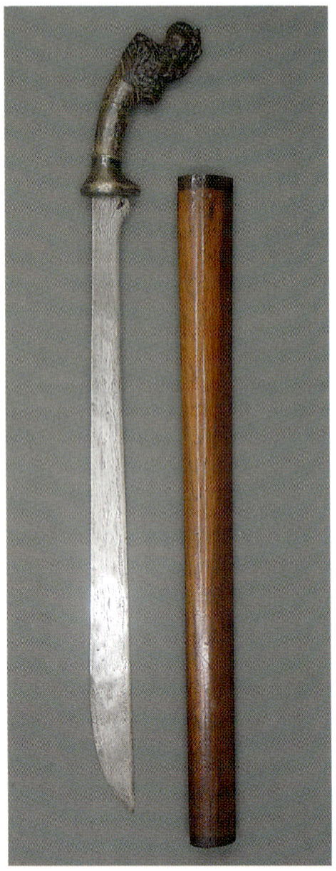

PEDANG III
SUMATRA, PALEMBANG

A sword usually with a slightly curved, sometimes straight blade. This blade may be forged smooth, with *pamor* or incrustations. Near the hilt (made of horn) or near the point forged ornaments may occur. The blade's sides are either flat or concave. The hilt's end has leaf-shaped motifs carved in relief. Usually, the hilt is silver plated, sometimes covering only the hilt's lower part, sometimes covering the entire hilt from the blade to the carved part. This plating may have chased or chiselled ornamental motifs. The scabbard may be made of smoothly finished wood, but often has silver bands, or is entirely silver plated.
(FISCHER 1918; MUSEUM 1965)

397. **Pedang III**
Sumatra, Palembang. L. 66 cm.

PEDANG ABEUSAH
SUMATRA

A type of sword.
(ROGERS)

PEDANG ACEH
See *rudus*

PEDANG BENTOK
SUMATRA, JAVA, BALI

A sword with a slightly curved blade with various forms. The hilt of this type of *pedang* does not have a hand protection.
(GARDNER 1936)

PEDANG BERANDAL
KALIMANTAN

A heavy type of *pedang*.
(GARDNER 1936)

PEDANG BERTUPAI

A sword in the shape of a Indian *tulwar* with a basket-hilt.
(GARDNER 1936)

PEDANG CHAKOK

A variant of the *pedang*.
(GARDNER 1936)

PEDANG CHEMBUL
BALI

A variant of the *pedang* with a basket-hilt.
(GARDNER 1936)

PEDANG DJAWIE BESAR
See *pedang jawie besar*

PEDANG ITEPA
See *pedang teupeh*

PEDANG JAWIE BESAR
[PEDANG DJAWIE BESAR]
KALIMANTAN

A sword with a straight blade with two sharp edges. The blade broadens towards the point. The hilt is straight with backwards curving protrusions.
(STONE)

PEDANG JENAWI
RIAU, MUAR

A long two-handed sword of Chinese or Japanese model, with a straight or curved blade. Both the Portugese and the Dutch employed Japanese mercenaries who left behind many weapons including swords. These were used by the local people because of the quality of the steel.
(GARDNER 1936)

PEDANG KASOQ
SUMATRA

A *pedang* with a straight, slightly flexible and double-edged blade. See also *kaso*.
(SNOUCK HURGRONJE 1892)

PEDANG LHEE KOERO
See *pedang lhee kuro*

PEDANG LHEE KURO
[PEDANG LHEE KOERO, PEDANG TOELOE KOEROE, PEDANG TULU KURU]

SUMATRA

A *pedang* with three grooves along the blade. Known as *pedang tuku kuru* (Gayo).
(KREEMER)

PEDANG LURUS
[PEDANG LUWUK]

JAVA

A variant of the *pedang* usually with a beautifully decorated silver hilt and scabbard. It has an almost straight, partly two-edged blade tapering towards the point. The hilt is round and near the end slightly curved. At the blade the hilt may have a broadened cross segment exactly fitting the scabbard with its broadened mouthpiece. Some examples have European blades.
(DIELES; ENGEL; GARDNER 1936; JESSUP; RAFFLES 1817)

398. **Pedang lurus**
Java.

400. **Pedang lurus**
Java. RMV 1848-3. Acq. in 1913 from dr. A.W. Pulle. L. 56.5 cm.

401. **Pedang lurus**
Java. L. 72.5 cm.

402. **Pedang lurus**
Java. L. 47.5 cm.

PEDANG LUWUK
See *pedang lurus*

PEDANG MENGERAT LEHER-LEHER
See *pedang pemanchong*

PEDANG MEUTAMPOE
See *pedang meutampu*

PEDANG MEUTAMPU
[PEDANG METAMPOE]

SUMATRA, ACEH

A *pedang* with a sabre-hilt ending in a knobbed decoration. Known as *pedang metampu* in Gayo.
(KREEMER)

PEDANG ON DJOQ
See *pedang on joq*

PEDANG ON JOQ
[PEDANG ON DJOQ]

SUMATRA

A *pedang* with a long, narrow blade.
(SNOUCK HURGRONJE 1817B)

PEDANG PAKPAK
See *podang*

PEDANG PEMANCHONG
[PEDANG MENGERAT LEHER-LEHER]

A sword once used for executions, on order of the ruler.
(GARDNER 1936)

PEDANG PERBAYANGAN
A variant of the *pedang*.
(GARDNER 1936)

PEDANG PEUSANGAN
See *sikin pasangan*

PEDANG RAJA OEDJONG
See *pedang raja ujong*

399. **Pedang lurus**
Java. A Javanese in war dress, carrying a spear, two *kerisses* and a *pedang lurus*.

PEDANG RAJA UJONG
[PEDANG RAJA OEDJONG, PEDANG TROEKI]

SUMATRA

A *pedang* with a blade which curves sharply from about half way along the blade.
(SNOUCK HURGRONJE 1892)

PEDANG SHAMSHIR
See *podang*

PEDANG SIEM

SUMATRA

A curved sword from Siam (Thailand).
(ROGERS)

PEDANG SOEDOEK
See *pedang suduk*

PEDANG SUDUK
[PEDANG SOEDOEK]

JAVA

A sabre with a hilt made of wood or horn in the shape of a *wayang* figure or a fancifully carved head. The blade may sometimes have *pamor*.
(HEIN 1899; JASPER 1904)

PEDANG TEUNOEANG
See *pedang teunuang*

PEDANG TEUNUANG
[PEDANG TEUNOEANG, PEDANG TOEANGAN, PEDANG TOEANGON, PEDANG TUANGON]

SUMATRA, ACEH

A *pedang* of cast iron. Also known as *pedang tuangon* (Gayo).
(KREEMER)

PEDANG TEUPEH
[PEDANG ITEPA]

SUMATRA, ACEH

A *pedang* of forged iron. Also known as *pedang itepa* (Gayo).
(KREEMER)

PEDANG TOEANGAN
See *pedang teunuang*

PEDANG TOEANGON
See *pedang teunuang*

PEDANG TOELOE KOEROE
See *pedang lhee kuro*

PEDANG TROEKI
See *pedang raja ujong*

PEDANG TUANGON
See *pedang teunuang*

PEDANG TULU KURU
See *pedang lhee kuro*

PEDEENG
See *pedang*

PEDENG
See *pedang*

PELADJU
See *pelaju*

PELAJU
[PELADJU, PERLADJU]

SUMATRA, BATAK

A ceremonial sword which the *datu* (magician-*cum*-priest) waves and makes gestures with during ceremonial dances. The blade has a straight back and an S-shaped edge. This edge has a slight broadening at the hilt. The hilt is of deer horn, slightly curved and ending in a point curving towards the blade's back. The hilt's foot has a decorated silver ring. The scabbard curves at the point towards the back and has at the end a crescent-shaped protrusion. Its halves are kept together by means of metal rings. See also *piso halasan*.
(MULLER; ROGERS)

PELANGKA
See *sikin pasangan*

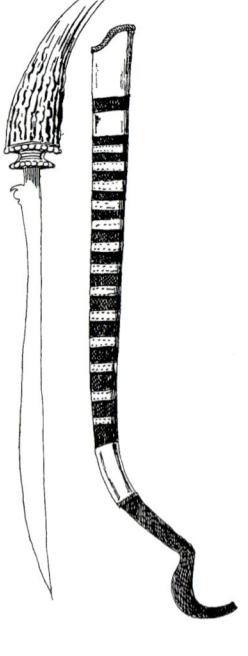

PELANTEK

A spear trap with a spring, located along the approaches to a fortification.
(GARDNER)

PENAI

SULAWESI, TORAJA

A machete of the Baree-speaking Toraja. Its blade broadens somewhat at the point. The edge is longer than the back, which turns in a slight curve towards the edge. The hilt is carved from buffalo horn and has a striking ornamentation. Most hilts are flattened and turn at a right angle half way up. Just past this turn we see a protruding ring around the hilt. At its end we normally see a V-shaped indentation and a number of cross ribs. At the blade the hilt has a broader part which fits precisely with the broader mouth of the scabbard. At the end of the scabbard we see a small foot with, resembling the hilt, a protruding ring. The scabbard is often decorated with beautiful carvings. Popular was a decoration with tin foil only meant for renowned head-hunters. The *penai* is carried with a belt around the waist. Therefore the scabbard has on the outer side a thickened protrusion in which two holes are made to run the belt through.
(DRAEGER; VOSKUIL 1921)

403. **Pelaju**
Sumatra, Batak.

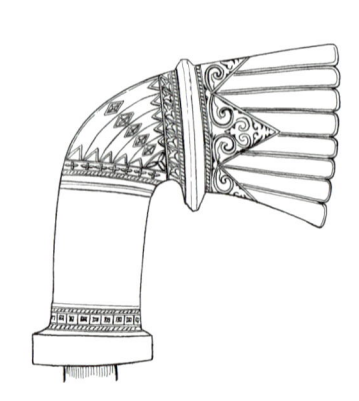

404. **Penai hilt**
Sulawesi, Toraja.

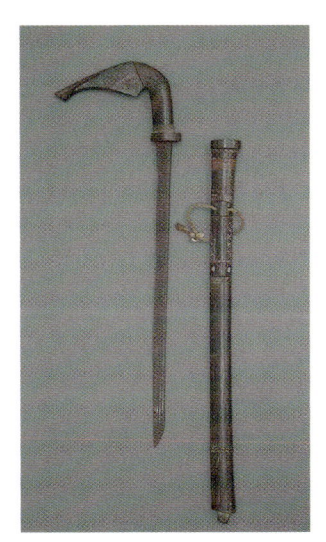

405. **Penai**
Sulawesi, Toraja. L. 56.5 cm.

406. **Penai**
Sulawesi, Toraja. L. 56.5 cm.

407. **Penai**
Sulawesi, Toraja. L. 67 cm.

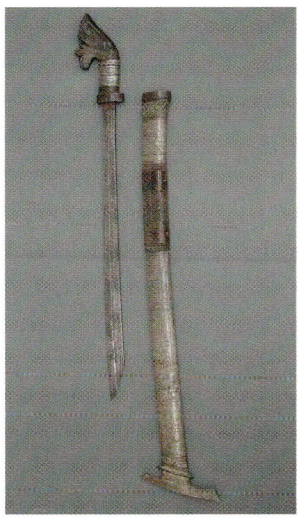

408. **Penai**
Sulawesi, Poso (?) L. 65.5 cm.

409. **Penai**
Sulawesi, Toraja. L. 60 cm.

410. **Penai**
Sulawesi, Toraja. L. 61.5 cm.

411. **Peurawot**
Sumatra, Aceh. Hilt made of *akar bahar* and ivory. L. 28 cm.

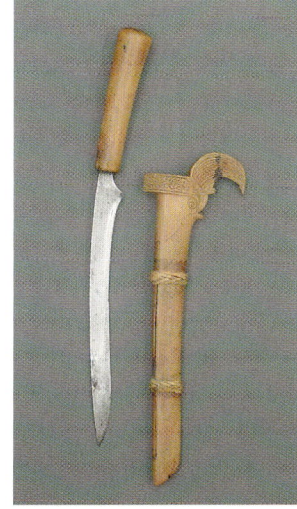

412. **Peurawot**
Sumatra, Aceh. L. 35 cm.

PENDAHAN
See *leumbeng*

PENUMBAK TEMBAGA
A brass knuckle-duster.
(GARDNER 1936)

PERISAI
See *peurise*

PERISE
See *peurise*

PERLADJU
See *pelaju*

PERMATA
See *bawar*

PERRISSE
See *peurise*

PETJUT
See *pecut*

PEUDEUENG
See *pedang*

PEUDEUENG PEUSANGAN
See *sikin pasangan*

PEULANGKAH
See *sikin pasangan*

PEUNOEWA
See *tumbok lada*

PEURAWOT
[PORAWET, SIKIN PEURAWOT, SIKIN RAWOT]

SUMATRA, ACEH, PIDIE
A *pinang* knife, sometimes with a golden or *suasa sampa* (decoration of the hilt near the blade) and *tampo* (knob of the hilt).
(VELTMAN; ROGERS)

PEURISE
[AMPANG AMPANG, PARINSE, PARISE, PARITSE, PERISAI, PERISE, PERRISSE, PRICEI, PRISE]
SUMATRA, ACEH, BATAK

A round shield with countless variations. On the inside we see several rings through which a rope or tape is pulled; through this, one places the left arm, holding on to part of the rope or tape with one's hand. The interior also has a small cushion to soften the blow on the shield. In Gayo (where the *peurise* is called *prise*) and Alas (where the *peurise* is called *ampang ampang*), the *peurise* has on the inside a rattan arm ring (*kela*) and a wooden grip (*amat amaton*). The wooden shields in Gayo (larger than those of Aceh) are covered with buffalo hide on the outside, sometimes even with a tiger skin. Characteristic of Aceh shields are the added semi-spherical knobs with a smooth or indented rim (*sikureueng dek* or *limong dek*) or star-shaped knobs of brass. Shields with four, five, six, seven or nine knobs are found. On the rattan shields (*peurise awe*) we see more elaborate knobs with a broad rim of à jour motif and a circle of *tumpals* (triangles) with stylised decorations inside them. If the decoration consists of seven stars, it imitates the *tuju bintang* (the Pleiades). The half moon is also found as a decorative motif. The *peurise* is carried on the back.

(FISCHER 1914; JACOBS; JASPER AND PIRNGADIE 1930; JESSUP; KREEMER; KRUIJT; VAN LANGEN; VAN DER TUUK)

PEURISE AWE
[PEURISE AWI]
SUMATRA, ACEH

A *peurise* of plaited rattan strips (*glong*). It may be covered with red or black cotton.

(JASPER AND PIRNGADIE 1930; JACOBS; KREEMER; STAAT)

PEURISE KAJEE
[PARITSE DONDANG, PEURISE DONDANG]
SUMATRA, ACEH

A wooden *peurise*.

(KREEMER; VAN DER TUUK)

PEURISE LEMBAGA
See *peurise teumaga*

PEURISE NILO
SUMATRA, ACEH

A *peurise* made of buffalo hide.

(KREEMER)

PEURISE PAROE
See *peurise paru*

PEURISE PARU
[PEURISE PAROE]
SUMATRA, ACEH

A *peurise* made of the skin of a sting-ray.

(KREEMER)

PEURISE TEUMAGA
[PEURISE LEMBAGA]
SUMATRA, ACEH

A *peurise* made of cast brass or bronze. The decoration often consists of concentric circles made on a lathe using a chisel.

(JACOBS; JASPER AND PIRNGADIE 1930; KREEMER)

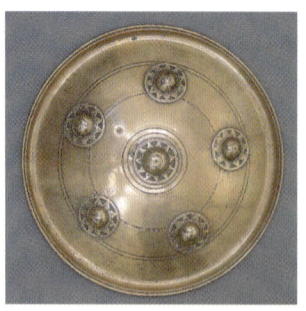

413. **Peurise awe** Sumatra, Aceh. Diam. 37 cm.
414. **Peurise awe** Sumatra, Aceh. Diam. 48 cm.
415. **Peurise awe** Sumatra, Aceh. Diam. 34 cm.
416. **Peurise awe** Sumatra, Aceh. Diam. 37.5 cm.
417. **Peurise teumaga** Sumatra, Aceh. Diam. 30 cm.
418. **Peurise teumaga** Sumatra, Aceh. Diam. 47 cm. 
419. **Peurise teumaga** Sumatra, Aceh. Diam. 25 cm. 
420. **Peurise teumaga** Sumatra, Aceh. Diam. 26 cm. 

PEURISE AWI
See *peurise awe*

PEURISE DONDANG
See *peurise kajee*

PINANG LAYAR
A bludgeon made of wood from *pinang liar*.

(GARDNER 1936)

PISAU
See *piso*

PISAU RAUT
See *piso raout*

PISAW
See *siraui*

PISO
[PISAU]

SUMATRA, ALAS

The people of Alas call all items used for cutting a *piso*, be it a weapon or a household knife.
(KREEMER)

PISO BELATI
JAVA, MADURA, SUMATRA, MINANGKABAU

A type of knife.
(DRAEGER)

PISO ECCAT
See *piso halasan*

PISO ENGKAT
SUMATRA, ACEH

A knife with a narrow blade measuring 10-25 cm. See also *piso halasan*.
(DRAEGER)

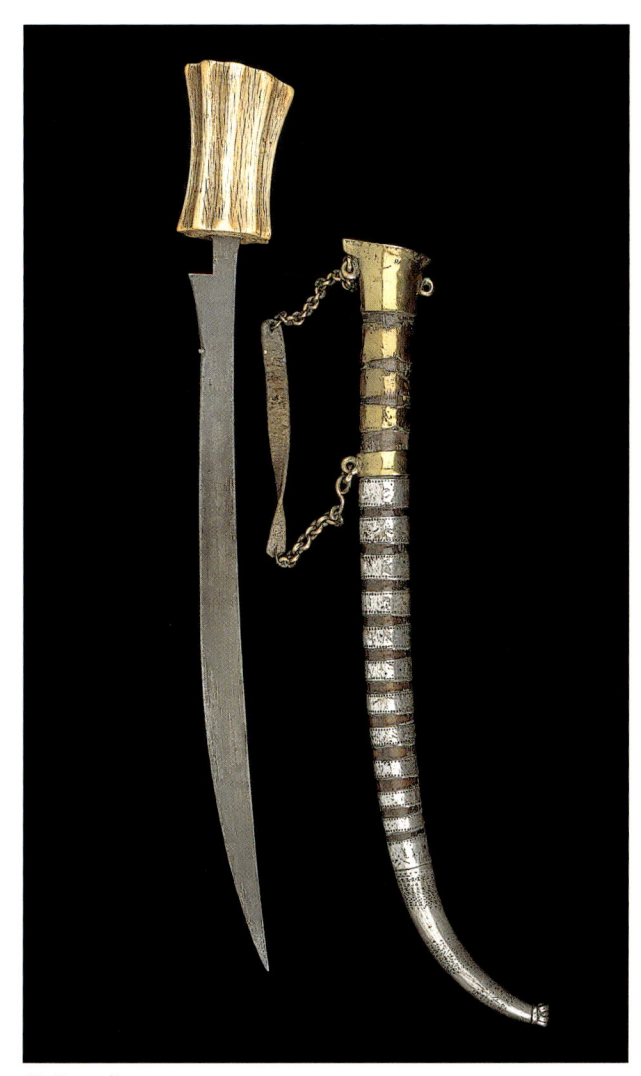

PISO GADING
SUMATRA, BATAK

A sword with a long, narrow blade ending in a sharp point. The back is slightly concave, the edge somewhat S-shaped. Near the hilt the blade is narrower and rectangular in cross-section. The hilt is short, thick and in the centre thinner forming a kind of hour-glass. It has deep grooves lengthwise and is often made of ivory or cast brass. The leather scabbard is thin, broader at the mouth and at the tip lightly and elegantly curved. If the scabbard is made of wood, it may be decorated, be covered with leather, or have narrow or broad metal strips. Brass scabbards are found, too. Their upper parts usually have a metal chain or belt.
(SIBETH; STONE; VOSKUIL)

PISO GULAK TAKA
See *kalasan*

PISO GULUK TAKA
See *kalasan*

421. Piso gading
Sumatra, Batak. RMV 3600-4432. Acq. in 1959 through the Ethnografisch Museum van de Koninklijke Militaire Academie, Breda. L. 61.5 cm.

422. Piso halasan
Sumatra, Toba Batak. L. 65 cm.

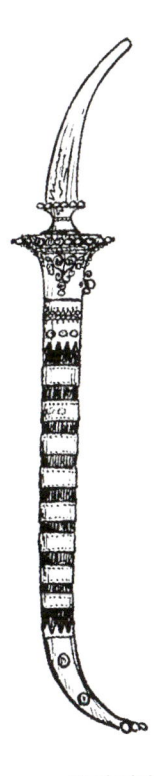

423. Piso halasan
Sumatra, Toba Batak.

PISO HALASAN
[ECCAT, EKKAT, ENGKAT, PISO ECCAT]

SUMATRA, BATAK

A sword with a hilt made of deer horn or antelope horn. Horn-shaped hilts of cast metal are also found. The blade has a straight back and a narrow point. Its broadest part is c.1/3 from the point. Towards the hilt the blade gradually becomes narrower. Near the hilt the blade suddenly broadens. The scabbard has a broader part at the top, usually ends at the bottom in a narrow curved point and is partly or totally covered with metal.
(ROGERS; SIBETH; STONE; VAN DER TUUK)

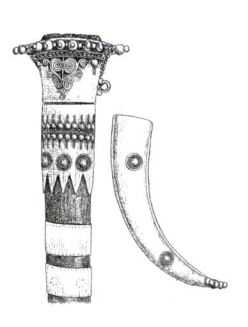

424. Piso halasan
Details of the scabbard of illus. 423.

PISO NI DATU
SUMATRA, BATAK
A knife with a straight-backed blade. Its edge is somewhat S-shaped and runs to the back at the point. The hilt is carved into a human figure with a broad, tapering metal ring at the bottom. The scabbard broadens at the opening and is cut in the shape of a mount. Whenever the knife is inserted into the scabbard, it looks as if the human figure is seated on the animal.
(VAN DER TUUK)

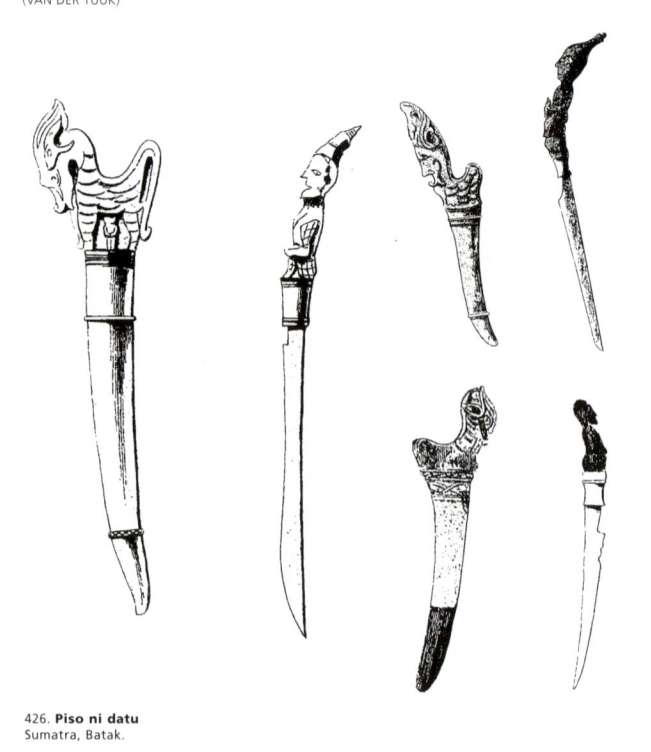

425. **Piso halasan**
Sumatra, Toba Batak. RMV 3155-17. Acq. in 1954 from W.G. Broek. L. 64 cm.

426. **Piso ni datu**
Sumatra, Batak.

PISO KALASEN
See *kalasan*

PISO KOERAMBI
See *korambi*

PISO KURAMBI
See *korambi*

PISO LAMPAKAN
SULAWESI, TORAJA
A knife, used as a tool, and as a weapon in case of danger.
(DRAEGER)

PISO LAPAN SAGI
See *sikin lapan sagu*

PISO MARIHOER
See *piso marihur*

PISO MARIHUR
[PISO MARIHOER]
SUMATRA, TOBA BATAK
A knife of a magician-*cum*-priest (*datu*). The term *piso marihur* literally means: 'tailed knife'.
(FISHER 1914)

PISO PERLADJO
See *piso perlajo*

PISO PERLAJO
[PISO PERLADJO]
SUMATRA, KARO BATAK
A knife used by a *guru*. The blade is slightly curved, back and convex edge run parallel. The back is concave and sharpened for c.1/3 of its length. The edge bends towards the back at the point. The hilt is octagonal and thinnest in the centre.
(FISCHER 1914)

PISO PODANG
See *podang*

PISO RAOET
See *piso raout*

PISO RAOUT
[HAUT NYU, JABANG, LANGGAI, LANGGEI, LUNGAT, MUNBAT, PISAU RAUT, PISO RAOET, POE]

A knife with a 5-10 cm long blade and a hilt measuring c.30 cm. On Kalimantan one of the blade's sides is usually convexly shaped. The other is concavely shaped, as is the case with *mandau*s. The blade is set into the hilt, obliquely to the handle and secured with resin (*damar*) and plaited rattan. The thin hilt is slightly curved, often undecorated and round in cross-section. Its end may have a beautifully carved figure and is often made of bone, (deer) horn or ivory.

The *piso raout* is found all over the archipelago. It is used for wood-carving and rattan-cutting (*piso*, meaning: 'knife'; *raout*, meaning: 'rattan'). When fine wood-carving is required, one holds the blade between thumb and index finger. The long hilt is held tight between the ribs and lower arm or under the armpit, using one's strength while at the same time working precisely.
On Kalimantan this type of knife is always carried with the *mandau* in a separate scabbard. This scabbard is made of palm leaf and attached to the back of the scabbard of the *mandau*. The Iban Dayak term for the *piso raout* is *munbat*, the Ngaju term is *langgei*, the Dayak of Baranjan say *jabang*. The Mendalam Kayan refer to it as *haut nyu*.
(AVÉ; HEIN 1890; HOSE; KREEMER; STONE; VOSKUIL 1921)

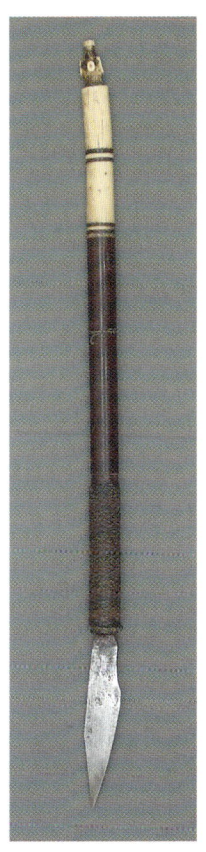

427. **Piso raout** Kalimantan. L. 42 cm.
428. **Piso raout** Kalimantan. L. 39 cm. 
429. **Piso raout** Kalimantan. L. 45.5 cm. 

PISO REMPU PIRAK
See *kalasan*

PISO ROEMBOE PIRAK
See *kalasan*

PISO RUMBU PIRAK
See *kalasan*

PISO SANALENGGAM
[PISO SANALENNGAN, PISO SINALENGGAM, PISO SINALENGGAN, PISO SURIK SINALENGGAN, SANALENGGAN]

SUMATRA, PAKPAK BATAK, TOBA BATAK
A sword with a broad blade. Its centre of gravity lies near the point. The edge is somewhat S-shaped and runs towards the back in a slight curve. The back is straight for 3/4 of the length and then runs at an obtuse angle to the point. The hilt may be carved in the shape of a squatting human figure or split at the end. The hilt's lower part may have a cylindrical metal sleeve. The scabbard is flat and broad. A rib runs along the entire length. The scabbard's mouth broadens out. The end of the scabbard runs in a curve either towards the blade's edge or towards the back. The sword's name is probably linked to the vermilion colour the scabbard sometimes has.
(FELDMAN 1990; FISCHER 1914; ROGERS; SIBETH; VAN DER TUUK)

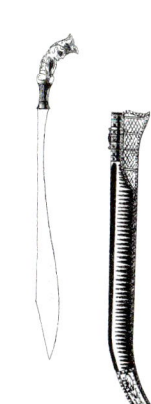

430. **Piso sanalenggam** Sumatra, Batak.

PISO SANALENGGAN
See *piso sanalenggam*

PISO SINALENGGAM
See *piso sanalenggam*

PISO SINALENGGAN
See *piso sanalenggam*

PISO SUKUL GADING
See *tumbok lada*

PISO SURIK SINALENGGAN
See *piso sanalenggam*

PISO TONGKENG
KALIMANTAN, DAYAK
A combat knife with a straight, rectangular blade. Its hilt, around which a thread is wound, stands at an obtuse angle to the blade.
(HEIN 1890; STONE)

PISOE
See *pisu*

PISU
[PISOE]

SULAWESI, BOMBA, BESOA
A stone thrower, made of wood, used mainly to scare birds away from the fields.
(GRUBAUER)

PLANGKAH
See *sikin pasangan*

PODANG
[PEDANG PAKPAK, PEDANG SHAMSHIR, PISO PODANG, PODONG, SHAMSHIR, TULWAR]

SUMATRA, BATAK
A long, slightly curved sabre often with a European blade. The name *podang* may well be derived from the Portugese *espadao*, pronounced *espandang*. Its back is concave, its edge (*mata ni podang* or *baba ni podang*) is convex. Sometimes *pamor* is found on these blades. The metal hilt has a cross-piece at the bottom to protect the hand and is more or less flattened. It broadens at each end, both roughly diamond-shaped or notched.
The hilt has a large bowl-shaped pommel to prevent it falling out of the hand. To make it as light as possible, this pommel is hollow. In the centre a small protrusion often occurs. A tuft of hair may be

431. **Piso sanalenggam** Sumatra, Batak. Hilt: *sukul nganga*. L. 57 cm.

fixed in the bowl. Hilts are usually made of brass, but may sometimes be of iron, silver or bronze. Except for iron examples, they are cast in one piece. The shape may be an imitation of Indian or Portugese swords. This especially goes for the two protrusions on both sides of the hilt parallel to the blade, which fit exactly in the spaces in the scabbard's upper part, as found in most examples. The scabbard is made of wood, the two parts are held together by metal or rattan strips. The name for this weapon is *(piso) podang* amongst Karo-, Toba- and Angkola Batak. The Asahan Batak call it *podong*.

(FISCHER 1914; GARDNER 1936; MÜLLER; SIBETH; VAN DER TUUK; STONE)

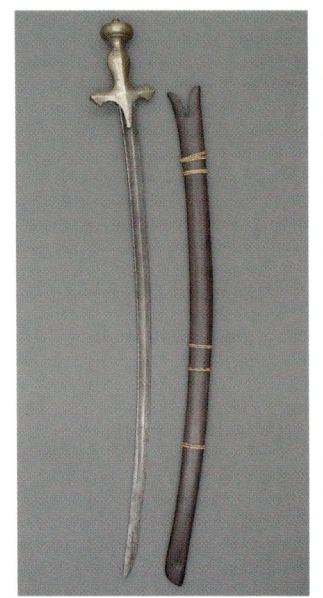

432. Podang
Sumatra, Batak. L. 71.5 cm.

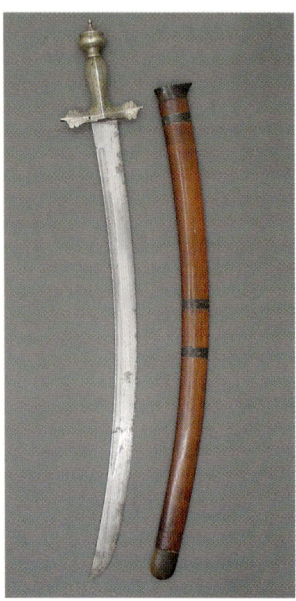

433. Podang
Sumatra, Batak. L. 72.5 cm.

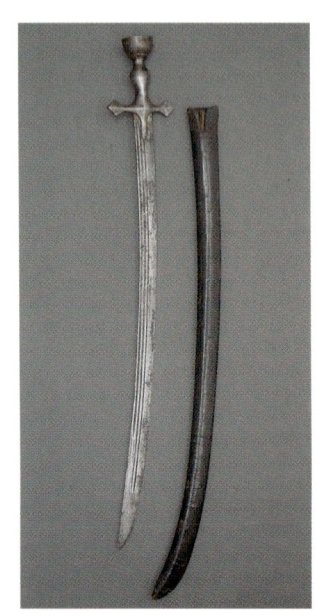

434. Podang
Sumatra, Batak. L. 92 cm.

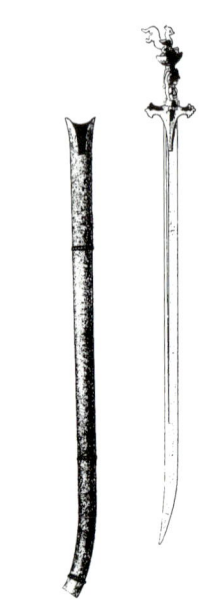

435. Podang
Sumatra, Batak.

436. Podang
Sumatra, Batak. RMV 3600-903. Acq. in 1959 through the Ethnografisch Museum van de Koninklijke Militaire Academie, Breda. L. 84.5 cm.

PODOENG
See *pedang*

PODONG
See *podang*

PODUENG
See *pedang*

POE
See *piso raout*

POEOT
See *puot*

POESSOE POESSOE
See *awola*

POKE
SULAWESI
A lance referred in the Makassar region to as *poke* and as *basi* in Buginese. The lances have various shapes and are named as follows: *lekonana*; *poke banrangang*, see *basi baranga*; *poke biring*; *poke laso tedong*; *poke mitoli*; *poke pammulu*; *poke pangka* (a royal weapon), see *basi paka*; *poke paroeng*, see *basi parung*; *poke parung*, see *basi parung*; *poke paso*; *poke-sangkoeng*, see *basi sangkung*; *poke sangkung*, see *basi sangkung*; *poke sipi*; *poke sipi nimangga*; *poke takang*, see *basi takang*; *poke-takkang*, see *basi takang*; *poke-tanjdjenang*, see *basi sanresang*; *poke tanjengang*, see *basi sanresang*; *tinjarung*.
(JASPER AND PIRNGADIE 1930; MATTHES 1874, 1885; SCHRÖDER)

POKI POKI
See *ekajo*

PORAWET
See *peurawot*

PRICAI KAYU
FLORES, MANGGARAI REGION
An oval or round wooden shield.
(DRAEGER)

PRICEI
See *peurise*

PRINGAPUS
BALI, SINGARAJA
A knife made of bamboo ending in a long sharp tip, sometimes smeared with poison.
(DRAEGER)

PRISE
See *peurise*

PRONGIN
See *utap*

PUE SURING
SUMATRA, BATAK
A machete with a hilt which has a large protrusion, branching off at an obtuse angle.
(HEIN 1899)

PUNGLU
BALI
A blow-pipe dart with a smooth tip smeared with poison. The other end has a small cone of pith.
(STONE)

PUOT
[POEOT]
KALIMANTAN, PUNAN
A blow-pipe made of polished wood with a horn ring on the mouth-piece. The *puot* may or may not have the point of a spear attached to the front.
(STONE)

PUSSU PUSSU
See *awola*

PUT
See *sumpitan*

437. "Raksasi" weapon carrier with ceremonial knife
Bali. RMV 1289-1. Acq. in 1901. H. 72 cm.

R r

ALPHABETICAL SURVEY

RAA
BANDA, KISSER PEOPLE
A sword with a scabbard (*kisa*).
(FOY)

RADJA DOEMPAK
See *raja dumpak*

RAJA DUMPAK
[RADJA DOEMPAK]
SUMATRA, ACEH
A sword with an S-shaped blade. A ceremonial weapon of foreign origin, very rarely found and quite similar to the *pedang*.
(VAN LANGEN; NIEUWENHUIZEN)

RANDO
See *randu*

RANDU
[BUNO, BUNO HAMPAK, RANDO]
KALIMANTAN
A tip of a ritual spear with two thin curved protrusions (*hampak*) and a thin metal shaft. The *randu* was used during festivities, such as death ceremonies (*tiwah*), in order to achieve 'mana', prosperity and to ward off evil. This weapon was no longer made after c.1890.
(HEIN 1890; SELLATO)

RANGGAS
A lance used during tournaments.
(GARDNER 1936)

RANGIN
A long shield covered with thin gold plate (*sarong perisai mas*): an indication of rank.
(GARDNER 1936)

RANJAU
[PANJI II, PANJIE, SUDAS, TUKAK]
A stick of *aren* wood, iron wood or split bamboo, one end of which is pushed into the ground in order to slow down attacks or pursuits. One type of *ranjau* measures c.10-15 cm and is aimed at injuring feet. A second, larger type is aimed at entering upper parts of the body. Both ends of the *ranjau* are pointed. The thicker one is stuck into the ground. The other end, pointing upwards, may be hardened by impregnating it with oil and smoking it over a fire. The *ranjaus* are pushed in the ground, well-hidden, sometimes upright, sometimes at an angle. The bamboo's rough fibres remain inside the wound causing irritation and infections, impeding recovery. In Sulawesi and Talaud, for example, *ranjaus* are treated with a poison called *tassen*. In Maluku *ranjaus* are made of a species of bamboo (*bulu tui*) which exudes a poisonous liquid. The Dayak term for *ranjau* is *tukak*.
(DRAEGER; GARDNER 1933; LOW; MARSDEN; ROTH 1896B; STONE; VOSKUIL)

RAOUT
See *rawet*

RAUT
SUMATRA, BATAK
A knife the blade of which has a straight back and an S-shaped edge. The type of hilt named *sukul jengkal bengkok* is found on the *raut*.
(VOLZ 1909)

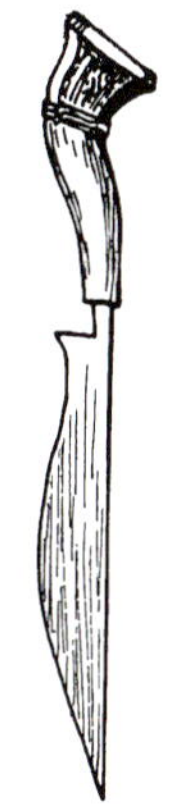

438. **Raut**
Sumatra, Karo Batak.

RAUT RAUT
See *bangkung II*

RAWET
[RAOUT]
SUMATRA, KARO BATAK
A small knife, serving various purposes such as splitting rattan.
(FISCHER 1914)

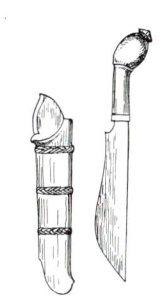

439. **Rawit** Sumatra, Batak.
440. **Rawit** Sumatra, Habinsaran, Batak. 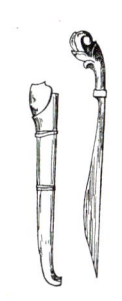
441. **Rawit** Sumatra, Kwalu, Batak. 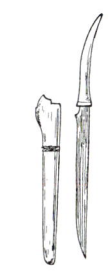

RAWIT
SUMATRA
A knife with a straight-backed blade and a more or less S-shaped edge. The blade varies in length and in breadth. The hilt has many different shapes.
(PARAVICINI; ROGERS; VOLZ 1909)

RAWIT PENGOEKIR
See *rawit pengukir*

RAWIT PENGUKIR
[RAWIT PENGOEKIR]
SUMATRA, BATAK
A double knife, mainly used for making lime holders.
(FISCHER 1914)

REMPU PIRAK
See *kalasan*

RENCHONG
See *rencong*

RENCONG
[RENCHONG, RENTJOENG, RENTJONG, REUNTJONG, RINTJOENG, RINTJONG, ROENTJAU]
SUMATRA, ACEH, GAYO, ALAS
A dagger not only found in Aceh, but also in Gayo and Alas. It has a sharp blade with a slightly convex back. Its edge is somewhat S-shaped because it curves towards the back at the tip, and due to a broadening (*duru seuke*) near the hilt. The blade has a strong back. Its sides (*gliba*) can be totally smooth, or partly cut out bringing about a slight concavity (*muka*) on both sides. This also causes slightly elevated ridges (*beuneung si urat*). Blades with a blood channel (*kuro*) along a part of the blade's length are also

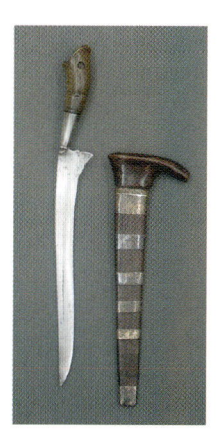

442. **Rawit** Sumatra, Batak. Hilt: *sukul tanke jambe*. L. 30.5 cm.

443. **Rawit pengukir** Sumatra, Batak. RMV 1767-47&48.

444. **Rencong** North Sumatra. L. 29 cm.

445. Rencong
North Sumatra. Hilt: *hulu puntung*. L. 36.5 cm.

446. Rencong
North Sumatra. Hilt: *hulu puntung*. L. 29 cm.

447. Rencong
Sumatra, Aceh. Hilt: *hulu puntung*. L. 36 cm.

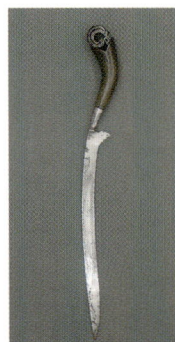

448. Rencong
North Sumatra. Hilt: *hulu puntung*. L. 42 cm.

449. Rencong
Sumatra, Aceh. Hilt: *hulu meucangge*. L. 38 cm.

450. Rencong
North Sumatra. Hilt: *hulu meucangge*. L. 33.5 cm.

451. Rencong
North Sumatra. Hilt: *hulu dandan*. L. 38.5 cm.

found. Where the blade enters the hilt, we see an octagonal stem-ring which has been forged on. On both sides hereof a small lip protrudes (*taku rungiet*). The blade often has a fine play of lines, due to the forging together of iron and steel. These flames or stripes (*kuree*) become more pronounced by the metal being exposed to the biting effect of lemon juice. Sometimes arsenic (*warang*) is added causing the metal to turn dark, almost black. Chasing and encrusting of blades also occurs. Blades from Gayo have a somewhat less elegant line. They are a little straighter and more pointed.

The hilt can be made of horn in various colours, elephant ivory, *akar bahar* (a species of black coral) or wood. Tusk of the *duyung* (sea-cow) or of seal and teeth of the sperm-whale are also found. Expensive *rencong* hilts are entirely or partially gold plated, finely carved, sometimes with filigree, sometimes with multicoloured enamel. In such cases the hilt's lowest part under the stem-ring is often decorated with double or triple *tampo*s or calyces. This stem-ring can be decorated with precious stones.

The following types of hilts are found:

(a) the *hulu meuccange*. This 'curved hilt' type has a thicker part in the middle called *boh mano* (meaning: 'chicken egg'). Above this part it curves in a right angle towards the blade's back. This curved part is usually round in cross-section broadening somewhat towards the tip, to end smoothly. The most popular material used for the *hulu meuccange* is horn. Hilts of ivory are two-pieced, the curved part is made separately.

(b) *hulu puntung*. This type has two variants: (1) with a blunt, rounded top which is sometimes flattened at the side and (2) with a shallow V-shaped indentation into which more or less open or closed leaf ornaments are carved.

(c) *hulu dandan*. A hilt made of 'white bone' (*i.e.*, tusk of, for instance, a sea-cow or *duyung*) and originating from Gayo. This hilt is round at the blade. At several centimetres from the blade, it makes a slight angle towards its back after which the hilt is thicker. A slight angle towards the same side follows, after which the hilt is thin and round in cross-section. This protruding part is much shorter than that of the *hulu meuccange*.

Most scabbards are made of wood and decorated with leaf patterns. In cross-section it is a flattened oval, following the blade's shape and usually coloured light brown. Its upper part broadens on both sides; more so on the edge's side, where it ends in a point facing downwards, than on the blade's back. The broadened upper parts are decorated with carved arabesques. Its lower part often has a protrusion decorated with carving, turning at an angle towards the blade's back. One or both sides of the whole scabbard are often decorated with carved figures. Scabbards can be two-pieced, or of one part which it is hollowed out through a channel cut along the longest part of the back. This channel is then covered with a very precisely cut piece of wood. The scabbard can also be totally or partially covered with a metal sleeve.

The *rencong* is carried on the left hand side, inserted between the body and girdle.

(BEIDATSCH; BISSELING; DRAEGER; EGERTON; FISCHER 1912; FORMAN; HEIN; GARDNER; JACOBS; JASPER AND PIRNGADIE 1930; KREEMER; KRUIJT; VAN LANGEN; MUSEUM 1965; STONE; VELTMAN; VOLZ 1909, 1912; VOSKUIL)

RENCONG PUCOQ PAKU
[RENTJONG POETJOQ PAKOE]

SUMATRA, ACEH

A *rencong* with an ornament attached to the hilt.
(SNOUCK HURGRONJE 1892)

RENTENG
KALIMANTAN

The term the Basi Sangiang use when referring to a spear.
(HEIN 1890)

RENTJOENG

See *rencong*

RENTJONG

See *rencong*

RENTJONG POETJOQ PAKOE

See *rencong pucoq paku*

REPAU
EAST FLORES

A spear measuring up to 3 m.
(DRAEGER)

REUNTJONG

See *rencong*

RINOMU

See under *o humaranga*

RINTJOENG

See *rencong*

RINTJONG

See *rencong*

RODA DEDALI
JAVA

A type of arrow used by deities, demigods and heroes.
(RAFFLES)

ROEDOES
See *rudus*

ROEDOES LENTI
See *ladieng*

ROEDOES LENTIK
See *ladieng*

ROEDOES PEH LAM TRIENG
See *rudus*

ROEDOES TARAH BADJI
See *tarah baju*

ROEDOIH
See *rudus*

ROEMBOE PIRAK
See *kalasan*

ROENTJAU
See *rencong*

RONKEPET
JAVA
A sword with a somewhat S-shaped blade. The hilt broadens at the blade.
(JASPER AND PIRNGADIE 1930)

RONTEGARI
JAVA
A sword the edge of which is slightly S-shaped. The back is straight for c.2/3 of its length, then curves sharply towards the edge, and continues in a slight curve towards the point. The part near the point therefore becomes lancet-shaped.
(JASPER AND PIRNGADIE 1930)

ROSO SEBUA
See *balato*

ROTI KALONG
JAVA
A knuckle-duster.
(DRAEGER)

RUDING LENGON
EAST JAVA
An ancient weapon with a heavy, fancifully shaped blade which may be thick along the back, ending in a point curving forwards. The edge is extremely S-shaped.
(GARDNER 1936)

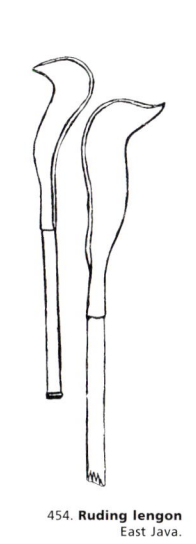

RUDOIH
See *rudus*

RUDOS
See *rudus*

452. **Ronkepet** Java.

453. **Rontegari** Java.

454. **Ruding lengon** East Java.

RUDUS
[CANDONG, CORIK, KLEWANG LAMTRIENG, KLEWANG LANGTRIENG, KLEWANG LIPEH OEDJONG, KLEWANG LIPEH UJONG, KLEWANG PEH LAM TRIENG, OEDJONG TIPIS, PEDANG ACEH, ROEDOES, ROEDOES PEH LAM TRIENG, ROEDOIH, RUDOIH, RUDOS, TJANDONG, TJORIK]

SUMATRA, ACEH, BATAK
A variant of the *klewang* which the Karo Batak call *rudus*. The Gayo call it *rudus* or *rudos*, the Pakpak call it *corik* or *candong*. The blade has a somewhat concavely curved edge. Its straight back curves towards the edge in a slight curve. The edge is sharpened from the point for c.2/3 or 3/4 of the blade. The back is sharpened from the point for c.1/3 of the blade. From this sharp part onwards the blade's sides are flat for a few centimetres. Next, the blade has a rib on both sides running closely alongside the blade's back to c.10 cm from the hilt. The ribs sometimes continue to the hilt, in that case the term *klewang lipeh ujong* is used. From the hilt onwards the back has several cross ribs decorated over a length of 20 cm with filed geometrical figures. The blade's edge also has, for several centimetres from the hilt, a number of cross ribs. The blade's length varies from 55-70 cm. The hilt is usually of the *hulu cangge gliwang* type (with protruding point) or of the *hulu tapa guda* type (without protruding point). The Pakpak are also familiar with hilts called *sukul jering* or *sukul ngangan*. Usually the *rudus* does not have a scabbard and is not carried in a belt, but in the hand. The blade is occasionally protected by means of a piece of goat's skin or palm-leaf. The *rudus* of the Pakpak, however, does have a scabbard. It is more a ceremonial weapon than meant for use in combat.
(DRAEGER; HEIN 1899; JACOBS; KREEMER; KRUYT; VAN LANGEN; MARSDEN; NIEUWENHUIZEN; ROGERS; SNOUCK HURGRONJE 1892; VERMEIREN; VOLZ 1909, 1912)

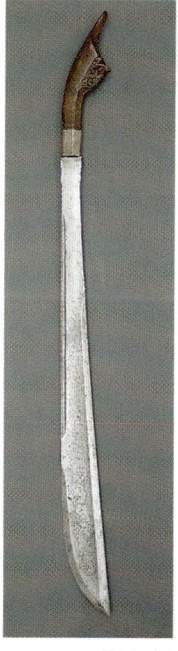

455. **Rudus** Sumatra, Aceh. Hilt: *hulu cangge gliwang*. L. 85 cm.

RUDUS LENTI
See *ladieng*

RUDUS TARAH BAJI
See *tarah baju*

RUGI
[RUGI GLAMANG]

NORTH SULAWESI, ALOR
An almost straight, narrow bladed sword. The back runs near the point in an angle towards the edge. At the point where the back curves a small barb occurs. The hilt ends in a large, carved triangular decoration. The scabbard has a large rectangular mouth-piece protruding only on the side of the edge. Its foot has a protrusion standing at an angle to the scabbard's remaining part, towards the edge. The scabbard is wound around with rattan over the entire length. The hilt and scabbard may be decorated with goat's and/or horse's hair. See also the entry: swords of the Timor group.
(FOY)

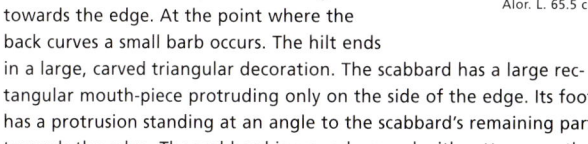

456. **Rugi** Alor. L. 65.5 cm.

RUGI GLAMANG
See *rugi*

RUMBU PIRAK
See *kalasan*

457. Shield
Sulawesi. RMV 1818-22. Acq. in 1912 through the Ethnografisch Museum van de Koninklijke Militaire Academie, Breda. L. 68 cm, W. 44 cm.

SABIET
See *sabit*

SABIT
[SABIET]

SUMATRA

A knife with a fanciful shaped, crescent-shaped blade found in various forms. It is used both as a weapon and as an agricultural tool. The hilt may have various forms, too, and is of about the same size as the blade. Variations of the *sabit* are the *padang sabit* and the *kampar sabit*.
(DRAEGER; STONE)

SABIT MATA DORA
KALIMANTAN

Two small curved knives, flat on one side and attached to each other at the top with a pin. They can be turned in such a way that the tips point in two directions and both hilts can be grasped in the centre.
(STONE)

SABOK
A band of silk or of fine cotton which is wound tightly around the body six or seven times, from the arm pits to the hips, like a military sash.
(GARDNER 1936)

SADAP
KALIMANTAN, MELANAU

A variant of the *parang latok* where the blade's shoulder is octagonal in cross-section.
(SHELFORD)

SADEB
See *sadeueb*

SADEB-SADEB
See *sadeueb*

SADEB TOENONG
See *sadeueb*

SADEUEB
[BAKONG, LADIING BOENGKOEWQ, LADIING PARAMBAH, PARANG BAKONG, SADEB, SADEB-SADEB, SADEB TOENONG, SADEUEB KOH EUMPEUENG KAMENG, SADEUEB KOH NALEUENG, SADEUEB TOENONG, SIKIN SADEUEB]

SUMATRA, ACEH

A weapon resembling the *ruding lengon*. Its heavy blade and hilt are approximately of the same length. The edge is narrow near the hilt, to then broaden in a curve. It is concavely curved near the point so that this point is at an angle of 90° to the blade. The back is mainly straight and bends near the tip in a quarter of a circle towards the point. The hilt is round and equally thick over the entire length.

The *sadeueb* is mainly used as a knife to cut grass with, but also may serve as a weapon. *Sadeueb koh naleueng* means 'grass-cutting *sadeueb*' and *sadeueb koh eumpeueng kameng* means: '*sadeueb* to cut goat fodder with'.
(JACOBS; VAN LANGEN; NIEUWENHUIZEN; ROGERS; SNOUCK HURGRONJE 1892)

458. **Sadeueb**
Sumatra, Aceh. L. 43.5 cm.

SADEUEB KOH EUMPEUENG KAMENG
See *sadeueb*

SADEUEB KOH NALEUENG
See *sadeueb*

SADEUEB TOENONG
See *sadeueb*

SADOEP
See *sadup*

SADOP
KALIMANTAN

A dagger with a short, broad blade which can be straight or slightly curved. The blade is double-edged and usually has an elevated rib in the centre. The hilt is symmetric and often decorated with *à jour* carving at the end.
(HOLSTEIN)

459. **Sadop**
Kalimantan. L. 26 cm.

460. **Sadop**
Kalimantan. L. 20 cm.

SADUP
[SADOEP]

KALIMANTAN

A knife with a long straight tapering blade, sharpened on one side.
(STONE)

SAGU SAGU
KALIMANTAN, ILANUN

A javelin used by the pirates of Ilanun.
(GARDNER 1936)

SAKIN
SUMATRA, MINANGKABAU

A knife with a short, straight blade.
(DRAEGER)

SALAWAKU
[EMULI, MA DADATOKO, SALUWAKU]

MALUKU, SERAM, AMBON, BURU, HALMAHERA, TOBELO

An hour-glass shaped, long shield. The *salawaku*, including the handle, is carved from a single piece of wood. The upper and lower part are broad, the shield is thinnest in the middle. At the front it is rounded or has a slight V-shape so that the centre part comes to the foreground. The shield is slightly curved from top to bottom. On the rear we see an elevated rib along the entire length, part of which is the handle in the middle. The front of the

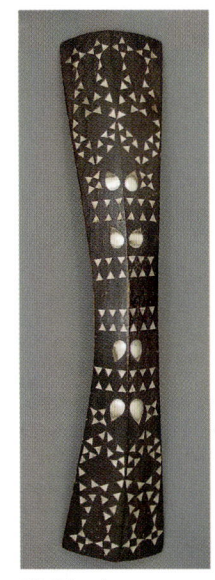

461. Salawaku
Maluku. Front view.
L. 69.5 cm.

462. Salawaku
Maluku. Side view.
L. 69.5 cm.

salawaku is painted black using soot and plant juice. It is inlaid with mother-of-pearl and fragments of earthenware, and/or painted with *kakean* symbols (used by a secret society) and other ornaments. These materials hint at the foreign aspect of the shields.
The term *salawaku* means 'protection' and 'repellence': a reference to the supernatural protection of the ancestors; or 'to miss and to catch'. With this technique the defender catches his adversary's weapon, secures it in the wooden shield and then disarms him. The shield forms a 'body' and the inlaid patterns refer to certain bodily parts. The upper segment refers to the head, the lower part to the feet. Arteries run lengthwise. The elevated rib on the rear side represents the spine and, just below the handle, the larynx. Inlays just above the centre represent the eyes. Their number allegedly refers to the number of enemies killed by the ancestors. The shield must be shorter than two arms length this in order to avoid the shield's end, held in place with a stretched arm in a horizontal position, reaching the chin and to avoid 'that tears fall on the shield'. Courage and sadness must be separated. The shield is not only a defensive weapon, but thanks to its peculiar, narrow shape may be easily moved to deliver blows with the sharp rims and corners. This shield is called *emuli* on the Buru Isles. The *salawaku* may be part and parcel of the bridegroom's marriage gifts and are worn during the *cakalele* (war dance) or the *hoyla* (war dance during the marriage ceremony among the Tobelo). During the *cakalele,* the *salawaku* is carried in the left hand.
(DRAEGER; FISCHER; HILKHUIJSEN; PLATENKAMP)

463. Salawaku
Seram. RMV 300-437. Acq. in 1878 World Fair, Paris.
L. 107 cm, W. 10 cm.

SALIGI
RIAU, LINGGA ARCHIPELAGO, BATAM, SULAWESI, CELATES AND BAJU PEOPLES
A plain spear with a wooden or bamboo shaft and a sharp point.
(DRAEGER)

SALOEKAT
See *salukat*

SALUKAT
[SALOEKAT]
MENTAWAI ISLANDS
A quiver made of a long segment of bamboo with a partition at the lower part and a long sharp protrusion at the tip.
(FISCHER 1909)

SALUWAKU
See *salawaku*

SAMAREMOE
See *amanremu*

SAMBILATIUNG
KALIMANTAN, KUALAKAPUAS
A type of spear.
(HEIN 1890)

SAMPAK GLIWANG
SUMATRA
A variant of the *klewang*.

SAMPIR
A broadened upper part of a scabbard.

SAMPULAU ANGGANG
KALIMANTAN, KENYAH
A battle head-dress made of rattan. It may be decorated with beadwork, feathers of the argus-pheasant or wood-cock, bits of coloured hair, hornbill beak and tail-feathers. See also *katapu*.
(AVÉ; COPPENS)

SANALENGGAN
See *piso sanalenggam*

SANG KAUW
A remarkably shaped weapon, used in the martial art called *kuntao*.
(DRAEGER)

SANGGA MARA
A forked weapon with barbs on the inside of both points, used to catch *amok* makers.
(GARDNER 1936)

SANGKOH
KALIMANTAN, SEA DAYAK
A spear with a long wooden shaft usually polished, sometimes with carvings. The point is c.30 cm long and broadest near the end. It is flat and attached to the shaft's side with rattan or brass wire. The shaft is usually made of iron-wood with a bone knob at the lower end. If this knob is absent, the shaft is finished in a point so that it may be stuck into the ground. It is held closer to the point than to the end, carried over the shoulder in the right hand (while the left hand holds a shield) with the tip pointing towards the ground. The *sangkoh* is used as a thrusting weapon at close quarters. Brass rings may indicate how often the owner has been on the war-path.
(HEIN 1890; ROTH 1896B; STONE)

SANGSANG
A tight jacket with buttons worn on top of the *sabok*.
(GARDNER 1936)

SANOKAT
MALUKU, SERAM
A spear, the most important weapon on Seram. It may have an iron or a wooden point.
(DRAEGER)

SAPITABON
KALIMANTAN
A type of spear.
(HEIN 1890)

SAPPOE
See *sapu*

SAPPU
See *sapu*

SAPU
[SAPPOE, SAPPU]
MID-SULAWESI, TORAJA
A bamboo blow-pipe containing a narrower tube. The term *sapu* is Buginese and Makassar.
(LOEBÉR 1928; MATTHES 1874, 1885; SCHRÖDER)

SAPURU
SULAWESI, UJUNG PADANG
A blow-pipe, used in days of old.
(DRAEGER)

SARAMPANG
[SERAMPANG, SEREPANG, SREPANG, TOEMBAK SERAMPANG]
KALIMANTAN
A spear used for catching fish and in combat. It has a bamboo shaft and may have between two and five metal barbed teeth. Its point is separate from the shaft, but tied to it with a length of string. Whenever a fish is caught, the point becomes loose from the shaft which, now serving as a float, remains connected to the point by means of the string preventing the fish from escaping.
(GARDNER 1936; JUYNBOLL 1909; ROTH 1896A; STONE)

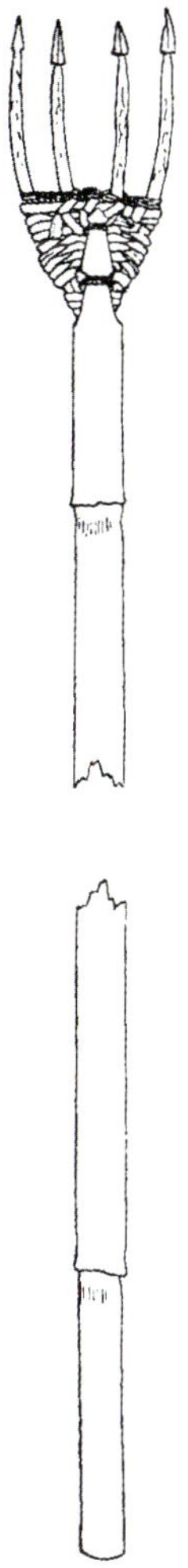

464. **Sarampang**
Kalimantan.

SAROBA
See *awola*

SEGU
JAVA
A metal, flexible staff with a horizontal protrusion, used to hit sensitive places on the arms and head.
(DRAEGER)

SEIMBOE NODA
See *seimbu noda*

SEIMBU NODA
[SEIMBOE NODA]
NIAS
A term referring to the scabbard of the *foda* (*balato*).
(FISCHER 1914)

SEIVA
See *sewar*

SEJWA
See *sewar*

SEKIN I
SUMATRA, PADANG HIGHLANDS
A dagger with a slightly curved blade. Its edge is concave, the back somewhat convex. On the edge's upper side, we see a protruding decoration of fine *à jour* work. The blade usually has ribs along its length. The hilt, shaped as a stylised bird's head, is first straight at the blade to then curve in an obtuse angle almost to a right angle towards the edge. The curved part is initially as thick as the rest of the hilt, but ends round and thin. The end is flat. The scabbard has at its mouth a decoration protruding towards the edge. The scabbard's end may be straight, or may be curved towards the back.
(FISCHER 1916; SNOUCK HURGRONJE 1892; STONE)

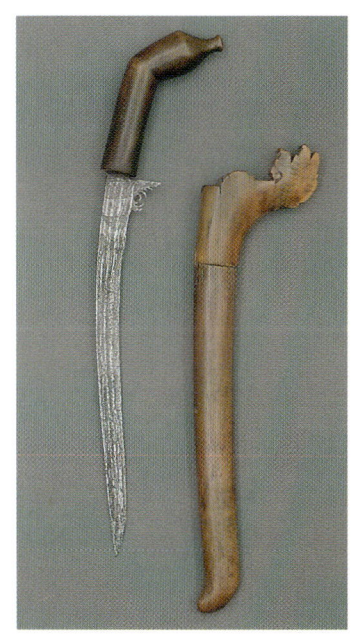

465. **Sekin**
Sumatra, Padang. L. 29.5 cm.

SEKIN II
SUMATRA, KARO BATAK
A machete, the standard Karo Batak working-*cum*-cutting knife. The blade has a straight back. Its edge is narrow along several centimetres at the hilt. It first goes in a concave curve and then in a convex curve ending in a broad part whereby the back and edge are almost parallel. Its end has an angle of 90° to the back and edge. The hilt is in the shape of a club, its upper part is flat. The scabbard consists of two parts kept together by means of rattan bands and is rounded at the bottom in a semi-circle. Its mouth broadens somewhat towards the blade's edge.
(FISCHER 1914)

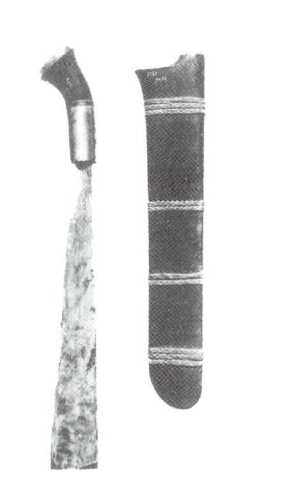

466. **Sekin II**
Sumatra, Karo Batak. RMV 1767-52. L. 42 cm.

467. **Sekin II**
Sumatra, Karo Batak. RMV 340-82. L. 34.5 cm.

SEKIN PANDJANG
See *sikin panjang*

SEKIN PANJANG
See *sikin panjang*

SEKIN PASANGAN
See *sikin pasangan*

SELIGI
A wooden arrow or spear varying from a light strip of bamboo sharpened on both sides (*seligi tajam bertimbal*), to a heavy, wooden javelin with a hard wooden point.
(GARDNER 1936)

SEMAREMOE
See *amanremu*

SENANGKAS BEDOK
A sabre with a heavy slightly curved blade with broad, shallow ribs. The hilt has a curled knob and no hand protector.
(STONE)

SEPA
JAVA
A sabre of which the blade is clearly S-shaped.
(DRAEGER; JASPER AND PIRNGADIE 1930)

SERAMPANG
See *sarampang*

SEREPANG
See *sarampang*

SERUNJONG
A pointed stick used as a pike.
(GARDNER 1936)

SEWAH
See *sewar*

SEWAH KECIL
[SEWAH KETJIL]
NIAS
A small knife.
(FISCHER 1909)

SEWAH KETJIL
See *sewah kecil*

SEWAR
[SEIVA, SEJWA, SEWAH, SIVA, SIVAS, SIWAH, SIWAI, SIWAIH, SIWAR, SIWAZ]
SUMATRA
The *sewar* is a dagger of Indian origin carried in a belt. The blade is either almost straight or slightly curved. Its length varies, from c.12 cm to such a length, specially in Central Sumatra, where it may serve as a machete, too. Its back is rather thick, the edge is located on its concave side and curves at the tip towards the back.

468. **Sepa**
Java.

469. **Sewar**
Sumatra, Minangkabau. L. 35.5 cm.

470. **Sewar**
Sumatra, Minangkabau. L. 30 cm.

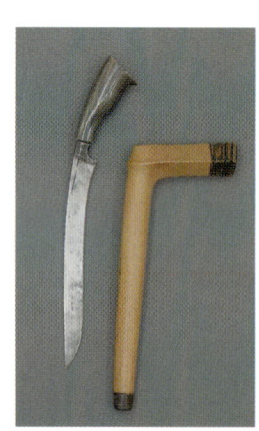

471. **Sewar**
Sumatra, Minangkabau. L. 20 cm.

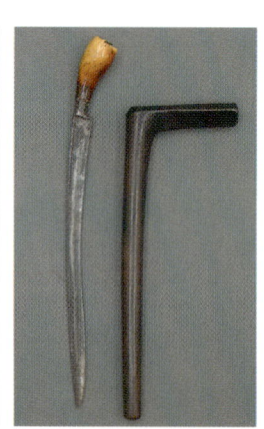

472. **Sewar**
Sumatra, Minangkabau. L. 33.5 cm.

473. **Sewar**
Sumatra. L. 32.5 cm.

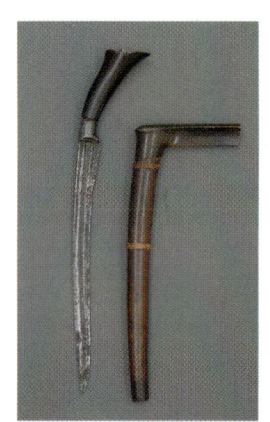

474. **Sewar**
Sumatra. L. 29.5 cm.

475. **Sewar**
Sumatra. L. 19.5 cm.

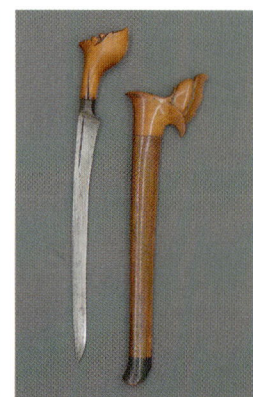

476. **Sewar**
Sumatra. L. 23 cm.

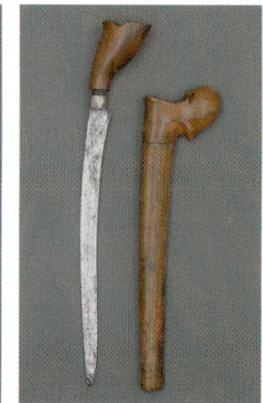

477. **Sewar**
Sumatra. L. 30.5 cm.

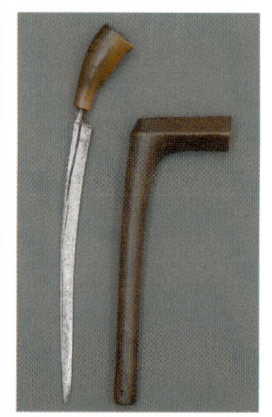

478. **Sewar**
Sumatra, Minangkabau. L. 28 cm.

479. Sewar
L. 24 cm.

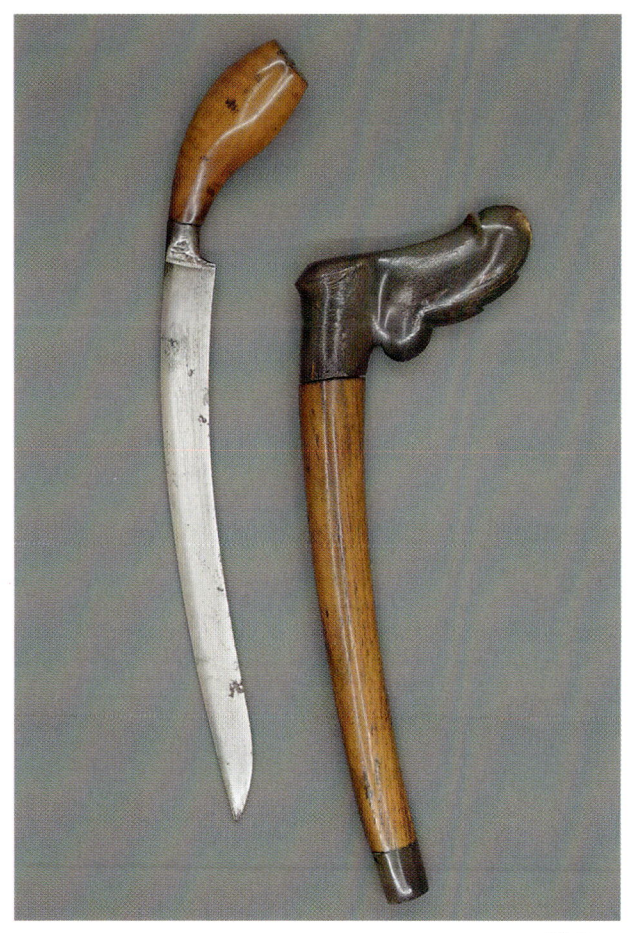

The blade often shows a seemingly haphazard, irregular groove along the back. It runs on both sides, near the hilt. Here the blade broadens somewhat to then continue in a forged-on stem-ring which is round, oval or angular in cross-section. If this stem-ring does not end in a cup-shape, the cross-section may be rectangular, hexagonal or octagonal. A cup-shaped or crown-shaped stem-ring usually has nine or sometimes eight sides. This ring then forms the transition between blade and hilt. Initially, the blade was made of (soft) iron. For the larger knives also used as machetes, harder bronze may be used. Only under the influence of European culture steel blades were produced.

Especially with the ceremonial *sewar*s much attention is paid to the stem-ring between blade and hilt. If this stem-ring (*tampo*) consists of one or more triangular seemingly interlocking wreaths, which form the hilt's lowest part, it is called *puco*. If these ends of the triangles do not end in a point, but are rounded, the ornament gets a cup-shaped appearance and is called *glupa* (a calyx or half a coconut shell to which the skin is still attached). Both the *puco* and the *glupa* are made of brass or *suasa*, sometimes of gold and usually beautifully enamelled.

The hilt itself has many variations including the *hulu boh glima* or *hulu glimo* (meaning: 'resembling a pomegranate'). It can be made of wood, horn, ivory (*gading*) or *akar bahar*. More expensive examples, often totally or partly covered with precious metal, may be decorated with enamel or small stones, be smoothly finished or decorated with carvings. On hilts with a flat upper part we sometimes see a golden decoration called *tampo* (like the stem-ring).

The scabbard is oval or round in cross-section and sometimes made of a single piece of wood or two parts glued together. These halves have their seam usually on the sharp side and the blade's back, but sometimes at the sides. They are often held together by rattan, silver, gold, or *suasa* bands or rings. The scabbard may (almost) entirely be covered with precious metal. On one side of its mouth we see a protrusion which, with more expensive examples, is covered with *à jour* figures worked in precious metal, filled with enamel. This protrusion (*cangge*) is usually smoothly flattened, sometimes lavishly decorated carvings are found.

The *cangge* has the following basic forms:
(a) straight and rectangular in cross-section whereby the planes of the square form an angle of 45° with the scabbard;
(b) also straight but with a slightly T-shaped cross-section;
(c) broader, to the sides somewhat flattened ending in a kind of knob.

The *sewar* may be found under various names in a large part of Sumatra. It strongly resembles the *tumbok lada*. The latter, however, has a somewhat differently shaped hilt and usually a heavier blade. Local names given to the *sewar* are: *siva* or *sewah* (Alas), *siwaih* (Aceh), *sewah* (Gayo) and *seiva* (Minangkabau).

(BEIDATSCH; DRAEGER; FISCHER; FORMAN; GARDNER; JASPER AND PIRNGADIE 1930; JESSUP; KREEMER; MARSDEN; MUSEUM 1965; VAN LANGEN; PARAVICINI; STONE; VELTMAN; VOLZ 1909, 1912; VOSKUIL)

SEWAR MEUKEURAWANG
SUMATRA, ACEH
A *sewar* with fine gold work on the scabbard.
(SNOUCK HURGRONJE 1892)

SHAMSHIR
See *podang*

SHIELD
Apart from those with specific names (and as such included in this survey), other shields are found in the entire archipelago, too. Their measurements, shapes, materials and decorations show a large variety. A number of these shields are described below:

(a) Amongst the Batak (Sumatra) large, curved combat shields with a single handle were used during the 19th century. They had black horse's hair on the right and left side, white horse's hair is attached to the upper part. Offering no protection against fire-arms, these shields rapidly disappeared from c.1900 on.
(VOLZ 1909)

(b) At Sidikalang (Sumatra) we see dance shields, covered with human skin and decorated with a stylised human face. When dancing, tassels made of feathers of a white cock are attached to the shield's upper part.
(VOLZ 1909)

480. Shield
Kalimantan, Segai.

482. Shield, rear side
Sumatra, Bobasan, Gayo. Diam. c. 50 cm.

485. Shield
Kalimantan, Dayak. Small flat bark shield with a cane rim, wooden handle and carved slip of wood along the middle of the front. L. 58 cm, W. 24 cm.

481. Shield
Sumatra, Sidikalang, Kepas. L. c. 60 cm, W. c. 12 cm.

483. Shield
Kalimantan, Sundayak. Front and rear view. Diam. 88 cm.

484. Shield
Kalimantan, Koti River.

486. Shield
Kalimantan. Cut out of solid wood. L. 58 cm, W. 22 cm.

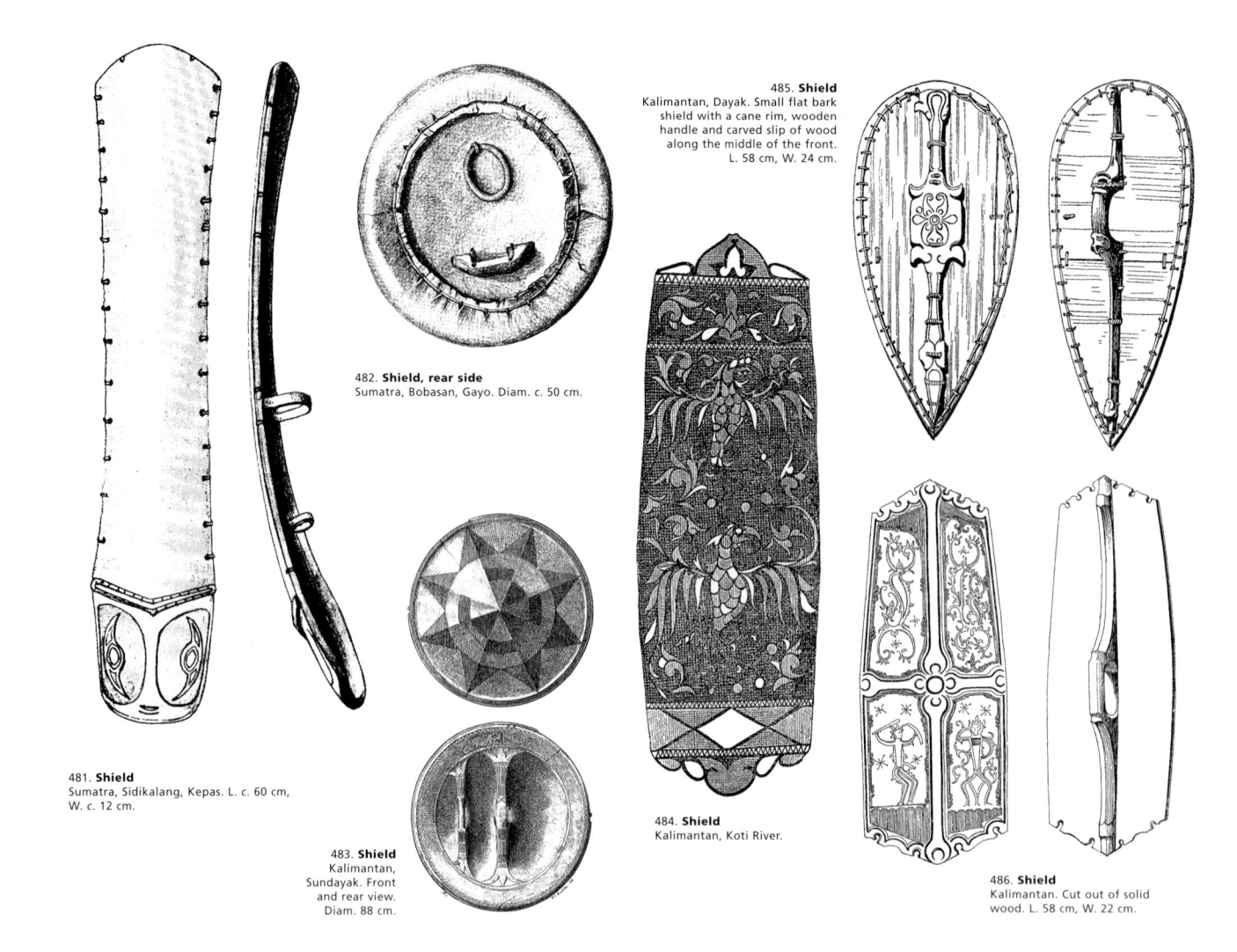

(c) As with Aceh shields, the Gayo have developed their own style: a round shield covered with buffalo hide and carried by means of a round loop and a handle on the lower arm, as is the case with the Batak. Offering no protection against fire-arms, these shields rapidly disappeared from c.1900 on.
(VOLZ 1912)

(d) On Enggano shields measured almost 2 m long and 1 m broad. Whenever fearing a raid the shields were used as a breast-work behind which the Engganese hid and from behind which they harassed the enemy with pikes.
(OUDEMANS)

(e) On Sumba round shields were made of plaited rattan and covered with buffalo hide.
(VOSKUIL 1921)

(f) Near the Koti River (Kalimantan) we find an extraordinary type of shield with a remarkable decoration.
(ROTH 1896B)

(g) The Land Dayak (Kalimantan) used shields made of tough bark with a rattan rim and a carved-out wooden board lengthwise in the middle of the shield. At its back runs a board out of which the grip has been carved lengthwise. This shield has an oval shape, the top is broader than its lower part. See also utap.
(HOSE; ROTH 1896A)

(h) The Dusun (Kalimantan) used a shield made of buffalo hide. They carried it on the lower arm attached to a belt, an unknown practice amongst other peoples.
(HOSE)

(i) In south-east Kalimantan the Segai had shields slightly curved lengthwise and very much curved in the width. They were decorated with a row of hair tassels near the upper and lower rims. Its upper and lower parts are straight.
(ROTH 1896A)

(j) The Sundayak of Kalimantan used a round shield made of solid wood with a conical cross-section. The front may have black-painted decorative motifs or may be undecorated. At the back two grips are carved, forming a whole with the shield. Handles made of rattan also occur. These shields seemingly belonged to the standard defence weapons of the Dusun.
(HEIN 1899)

(k) A wooden shield and handle cut from a single piece of wood, found on Kalimantan.
(ROTH 1896B)

(l) Characteristic of the Matana Lake region (Sulawesi), inhabited by the Tobelo, are long, slender shields made of wood or plaited rattan. Their cross-section shows a deep V-shape. These shields taper somewhat towards the lower and upper side. As far as shape goes, they resemble the *kantas* of the Poso region, which have an extravagant decoration of, sometimes variegated,

487. Shield
Sulawesi, Tobela.

488. Shield
Sulawesi, Tobela.

489. Shield
Sulawesi, Tolambatu.

490. Shield
Sulawesi, Tolambatu.

491. Shield
Sulawesi, Tolambatu.

492. Shield
Sulawesi. RMV 1818-22. Acq. in 1912 through the Ethnografisch Museum van de Koninklijke Militaire Academie, Breda. L. 68 cm, W. 44 cm.

493. Shield
Sulawesi, Toraja.

494. Shield
Sulawesi, Toraja.

495. Shield
Sulawesi, Toraja.

496. Shield
Sulawesi, Toraja.

coloured goat's hair, pieces of shell and bone.
(GRUBAUER)

(m) Shields found in the Tolambatu region (Sulawesi) are long, narrower at the centre, and broader at the top and bottom. This shield's cross-section shows a V-shape through which lengthwise along the middle a rib comes into being. On the back, we see the handle. A slightly protruding thicker point is located on the front opposite this grip.
(GRUBAUER)

(n) Shields of the Toraja living in Bilalang (Sulawesi) are usually made of leather, sometimes of wood. Their shapes are rectangular or somewhat tapering. They may be lavishly decorated with various geometrical patterns.
(GRUBAUER; RODGERS)

SI EULI

NIAS

A dagger with a narrow, straight blade carried diagonally in the centre of the belt. The hilt is separated from the blade by a cylindrical brass ring and is curved at the end or makes a slight curve at about halfway. In the latter case the top of the hilt is flattened. The scabbard is straight and has a cross-piece at the mouth protruding towards the blade's edge or towards both sides. To the rear it may have a small protrusion, but also a prominent protrusion the point of which curves somewhat upwards. The scabbard may be wound with a brass wire and may have a small angled foot. Sometimes it has small chains with bells.
(FELDMAN 1990; KOL)

497. Si euli
North Nias. Warrior with a *si euli* (dagger) in his belt and a *baluse* (shield).

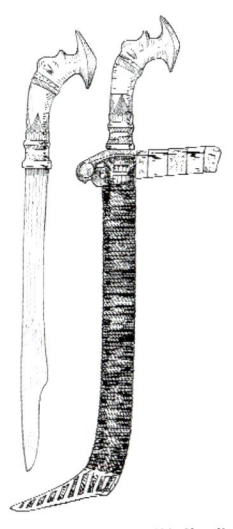

498. Si euli
Nias.

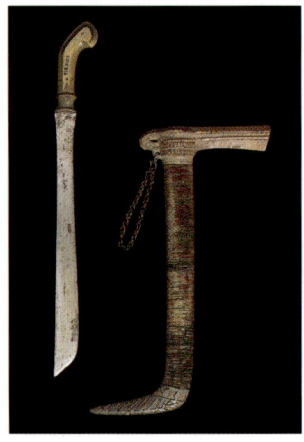

499. Si euli
Nias. RMV 718-25. Acq. in 1889 from M.J. Kleijer. L. 38.5 cm.

500. Si euli
Nias. L. 38.5 cm.

SI OR
SUMATRA, TOBA BATAK
A bow with which to shoot pellets. The small bag in which these pellets are stored is tied to the tip of the bow.
(STONE)

SIBAK
A type of knife or sword.
(STONE)

SIKAPAN
A jacket worn on top of the *sabok*, the *sangsang* and the *kotau*.
(GARDNER 1936)

SIKIM
See *sikin*

SIKIM GAJAH
[SIKIM GALA]
SUMATRA, NIAS, MENTAWAI ISLANDS, RIAU ARCHIPELAGO
A sword with a blade measuring c.50 cm, with a straight back and an S-shaped edge. The *sikim gajah* is carried in a scabbard.
(DRAEGER; GARDNER 1936)

SIKIM GALA
See *sikim gajah*

SIKIM PANDJANG
See *sikin panjang*

SIKIN ALANG
See *luju alang*

SIKIN DELAPAN SAGI
See *sikin lapan sagu*

SIKIN DELAPAN SAGOI
See *sikin lapan sagu*

SIKIN IKOE MANOK
See *sikin iku manoq*

SIKIN IKOE MANOQ
See *sikin iku manoq*

SIKIN IKU MANOQ
[SIKIN IKOE MANOK, SIKIN IKOE MANOQ]
SUMATRA, ACEH
A variant of the *sadeueb*. Its blade has the same width over the entire length.
(SNOUCK HURGRONJE 1904; ROGERS)

SIKIN LAPAN SAGOE
See *sikin lapan sagu*

SIKIN LAPAN SAGU
[LAPAN SAGI, LOEDJOE LAPAN SAGI, LUJU LAPAN SAGI, PISO LAPAN SAGI, SIKIN DELAPAN SAGI, SIKIN DELAPAN SAGOI, SIKIN LAPAN SAGOE]
SUMATRA, ACEH, GAYO, ALAS
This dagger called *sikin lapan sagu* (Gayo: *luju lapan sagi*; Alas: *piso lapan sagi*) is an *adat* weapon, seldom found in Aceh, but more often in the Gayo and the Alas regions. The blade's shape resembles that of the *sikin panjang*, of which both the edge and back are straight and parallel. The edge curves towards the back at the point. The hilt is referred to as *hulu lapan sagu* (Gayo: *hulu lapan sagi*; Alas: *sukul lapan sagi*) meaning: 'octagonal hilt'. It is made of *suasa* or gold, filled with resin, curved in the centre to end in two forked protrusions. The scabbard follows the blade's shape and has two protrusions at its mouth.
(KREEMER; NIEUWENHUIZEN; SNOUCK HURGRONJE 1904; VELTMAN; VOLZ 1909, 1912)

SIKIN MEUKSAROEEK OELAT
See *sikin meuksaruek ulat*

SIKIN MEUKSARUEK ULAT
[SIKIN MEUKSAROEEK OELAT]
SUMATRA
A *sikin panjang* with a type of hilt named *hulu peudada*.
(ROGERS)

SIKIN PANDJANG
See *sikin panjang*

SIKIN PANJANG
[ANDAR, GLOEPAK SIKIN, GLUPAK SIKIN, JEKINPANDJANG, LOEDJOE ACEH, LOEDJOE ATJEH, LOEDJOE NAROE, LOEDJOE NARU, LUDJU NARU, NARUMO, SEKIN PANDJANG, SEKIN PANJANG, SIKIM PANDJANG, SIKIN PANDJANG, SIKIN PANJANG MEUTATAH, THIKIN PANJANG]
NORTH SUMATRA
The *sikin panjang* (meaning: 'long *sikin*') is the most popular fighting weapon of the inhabitants of north Sumatra. In the early years of the Aceh War against the Dutch (which began in 1873 and lasted for over thirty years) many *sikins* were made, especially prior to 1879 when a start was made with the disarmament of the population. The spread of the *sikin panjang* is limited to Sumatra, and especially to Aceh and Gayo (where the term *luju naru* is used), but also in Alas (where it is named *andar*) and to a lesser degree in the Batak area.
Characteristic for the *sikin panjang* is the blade's shape. Similar blades only occur with the *luju alang* and with the daggers *sikin lapan sagu* and the *lopah petawaran*. The weapon is always provided with a scabbard, and is carried in the belt. The *sikin panjang* has the following main segments: blade (*wilah*), hilt (*hulu*), scab-

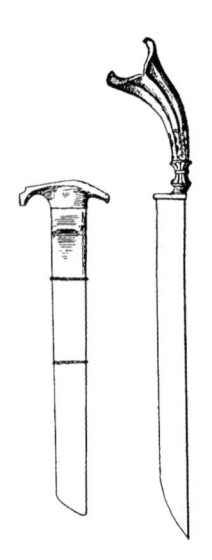

501. Sikin lapan sagu
North Sumatra. Hilt: *hulu lapan sagu*.

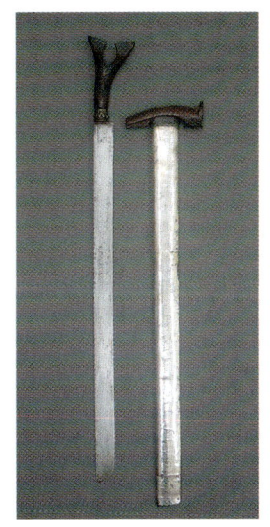

502. Sikin panjang
North Sumatra, Peusangan. Hilt: *hulu peusangan*. L. 78.5 cm.

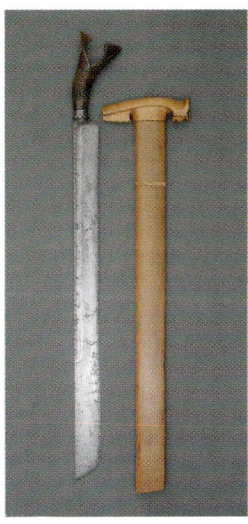

503. Sikin panjang
North Sumatra, Peusangan. Hilt: *hulu peusangan*. L. 66 cm.

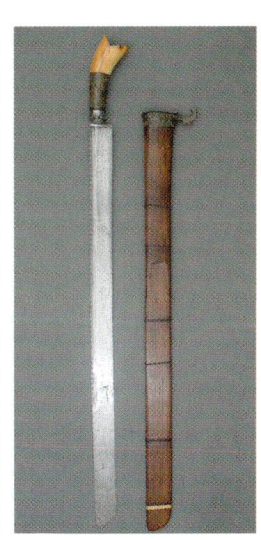

504. Sikin panjang
North Sumatra. Hilt: *hulu rumpung*. L. 77 cm.

505. Sikin panjang
North Sumatra. Hilt: *hulu tumpang beunteueng*. L. 70 cm.

506. Sikin panjang
North Sumatra. Hilt: *hulu tumpang beunteueng*. L. 73.5 cm.

bard (*sarung*). The blade is completely straight whereby the edge and the back run parallel. Near the tip the edge curves off towards the back. Along the back from the hilt onwards for c.2/3 of the blade, we find on both sides a broad, rather shallow groove. The cross-section of the blade is wedge-shaped. At the hilt we see a stem-ring forged in one piece with the blade. This stem-ring has eight or sometimes nine angles (*sepals*). The tang (the projection at the base of the blade) is also forged in one piece with the blade, and is glued into the hilt with resin. Sometimes the blade may be damascened (*kuree*) giving it a flamed, veined appearance. These veins (*reuta*) are considered to be auspicious or inauspicious signs by cognoscenti. The blade sometimes has gold-wire incrustations. Such a weapon is called *sikin panjang meutatah*.

The parts of the blade are:
- its sharp side (*mata*);
- its back (*rueng*);
- its flat sides (*gliba*);
- the grooves running lengthwise along the back (*kuro*);
- the part of the blade near the hilt; plated with gold, silver or *suasa*, it is called *sampa*;
- the point of the blade which is curved along the edge (*bungko*);
- the tongue (*puteng*).

The hilt which is always forked or mouth-shaped can be divided into three main shapes:
(a) broadening at the top (*hulu peusangan*). From its forked end the points broaden on the inside so that they come together somewhat, almost touching each other;
(b) with forked points showing a shallow V-shape; this is the most frequent type of hilt and is called *hulu rumpung*;
(c) with a deep V-shaped end, called *hulu tumpang beunteueng*;.

The material of which the hilt is made is mostly buffalo horn (*lungkee*). Sometimes of elephant tusks (*gadeng*) or the teeth of the sperm-whale. A combination of these materials also occurs. It may be entirely or partially covered with gold, silver or *suasa* and can be smoothly finished. It often has either a plain pattern of ribs, or shallow beautifully incised figures.

The hilt may also have one or more of the following brass, golden, silver or *suasa* decorations:
(a) the *reukueng rungiet* (the breast shield of a Capricorn beetle), a decoration covering the connecting part between the hilt and blade;
(b) the *glupa* (calyx), an ornament attached to the hilt's base as a ring or husk. It consists of one or more garlands of triangles, the points of which are rounded;
(c) the *puco* is a decoration resembling the *glupa*. With the *puco*, however, the triangle's tips are pointed. Both the *glupa* and the *puco* often show fine enamel work;
(d) the *saruek ulat* (meaning: 'cocoon of a caterpillar'), a thick elaborated husk which can be attached to the hilt's base instead of the *glupa* or *puco*.

The wooden scabbard is straight and shows at the point the same curve as the blade. Its cross-section is flat-oval. The part where the blade's back is located is a little broader. This scabbard can be made of a single piece, or of two pieces placed against each other and glued together. It is held together by rings or strips of silver, *suasa* or plaited rattan. It can be hollowed out through a narrow opening along the entire length of the back. This opening is then closed with a small piece of wood of the same type of which the scabbard is made.

The scabbard's upper part may be shaped in different ways. It may be cut from one piece along with the remaining part of the scabbard, or it may be a separate mouth-piece (*jambang*). In the last case, the mouth-piece is made from wood, horn or ivory. If cut from one piece along with the remaining part of the scabbard, it broadens in all directions. The rear side shows a short protrusion. The sharp edge's side has a longer, more or less pointed, curved protrusion (*cangge*). Quite often the entire upper part is decorated with carved arabesques and flower or plant motifs.

Another type of mouth-piece is large and heavy, stands at an angle of 90° to the scabbard and protrudes to both sides. Towards the blade's back this protrusion is a little longer and narrower, while shorter and blunter on the sharp side.

Such mouth-pieces may also be beautifully decorated, sometimes by filling the carvings with black resin. They are usually combined with a hilt of the *hulu peusangan* type, and is characteristic for the region around Lake Tawar, and the neighbouring parts of northern Aceh, especially Peusangan.

The scabbard's sides may be smooth or decorated, with exquisite carvings, arabesques or floral motifs, or with shallow, blackened incrustations. Scabbards completely covered with metal (*salob*) also occur. See also *galaijang tokong*.

(BISSELING; DIELES; GARDNER 1936; HEIN 1899; JACOBS; JASPER AND PIRNGADIE 1930; KREEMER; KRUIJT; VAN LANGEN; MUSEUM 1965; NIEUWENHUIZEN; VAN DER TUUK; VELTMAN; VERMEIREN; VOLZ 1912; VOSKUIL 1921)

SIKIN PANJANG MEUTATAH
See *sikin panjang*

SIKIN PASANGAN
[BEULANGKAH, BOLANGKAH, KLANGKAH, KOELANGKAH, KOELANGKAIH, LADING, PALANKA, PEDANG PEUSANGAN, PELANGKA, PEUDEUENG PEUSANGAN, PEULANGKAH, PLANGKAH, SEKIN PASANGAN]

SUMATRA, ACEH
A sabre with a curved blade with parallel edges, becoming narrower at the point. The blade has a stem-ring and is usually thin and flexible. However, thicker and non-flexible blades also occur. The *sikin pasangan* has a hilt of the *hulu tumpang beunteueng* type, a slightly curved hilt made of horn ending in the shape of a mouth, or a *hulu peusangan* where the points of the forked end almost come together. The *sikin pasangan* may be seen as the curved form of the *sikin panjang* or as intermediary between the *sikin panjang* and the *pedang I*.

(JACOBS; KREEMER; NIEUWENHUIZEN; ROGERS; SNOUCK HURGRONJE 1904; VERMEIREN 1987; VOLZ 1912)

507. Sikin pasangan
North Sumatra. Hilt: *hulu peusangan*. L. 95 cm.

508. Sikin pasangan
Sumatra, Aceh. Hilt: *hulu peusangan*. L. 86.5 cm.

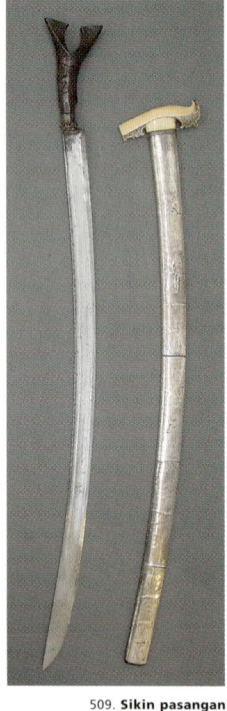

509. Sikin pasangan
Sumatra, Aceh. Hilt: *hulu peusangan*. L. 92.5 cm.

SIKIN PEURAWOT
See *peurawot*

SIKIN PREUNGGI
SUMATRA
A type of knife.
(ROGERS)

SIKIN RAWOT
See *peurawot*

SIKIN SADEUEB
See *sadeueb*

SILIGIS
KALIMANTAN
A wooden javelin.
(STONE)

SIMALA
See *sumara*

SIMOENOENG
See *simunung*

SIMONG
See *gagong*

SIMUNUNG
[SIMOENOENG]
SUMATRA
A type of sword.
(ROGERS)

SINA PAYED
KALIMANTAN, SARAWAK, BIDAYUH
A knife with a long round wooden hilt and a short blade. The hilt has near the blade a slight bend. The blade has a straight edge and back. Its back runs in an oblique angle towards the edge thus ending in a sharp point. The knife is used, for instance, to split rattan and bamboo and to make exquisite carvings in wood or bamboo. Except for a slight bend in the hilt, the *sina payed* is almost identical to the *piso raout*.
(CHIN)

SIPET
See *sumpitan*

SIRAUI
[PISAW]
SUMATRA, MINANGKABAU
A dagger with a sturdy, curved blade. The edge is located on the outside. The blade has one or more blood channels. The hilt is made of wood or horn, which smoothly follows the blade's curve to end in a thick, rounded upper part. The wooden scabbard is usually cut from one piece. At the mouth on the edge's side, one more or less elaborate decoration protrudes, more elaborate ones are in the shape of leaf or tendril motifs.
(FISCHER 1918)

510. Siraui
Sumatra, Minangkabau. L. 30.5 cm.

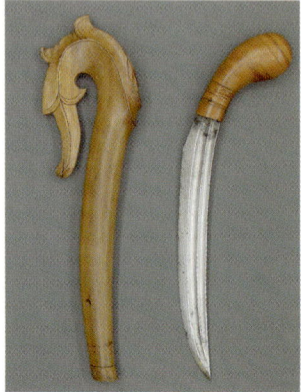

511. Siraui
Sumatra, Minangkabau. L.. 24.5 cm.

SIREN
See *ipoh*

SIVA
See *sewar*

SIVAS
See *sewar*

SIWAH
See *sewar*

SIWAI
See *sewar*

SIWAIH
See *sewar*

SIWAR
See *sewar*

SIWAZ
See *sewar*

SLIGHI
KALIMANTAN, SEA DAYAK
A javelin made mostly of iron-wood (*bilian*). Palm-wood, especially *imbery*, is also used. Its point is hardened in a fire. The enemy is tormented with a rain of such javelins before being close enough to be struck with normal spears.
(ROTH 1896B; STONE)

SOAT
[SOSOAT]
MENTAWAI ISLANDS
A spear with a flat iron point, with or without barbs. The shaft's lower part ends in a point.
(FISCHER 1909)

SODAK
A spear-like weapon with a broad blade.
(GARDNER 1936)

SOEDOEK
See *suduk*

SOEHOEL
See *sukul* or *hulu*

SOEKOEL
See *sukul* or *hulu*

SOEKOEL DJERRING
See *sukul jering*

SOEKOEL GERPONG
See *hulu puntung*

SOEKOEL LAPAN SAGI
See *hulu lapan sagu*

SOEKOEL MEKEPIT
See *hulu meu apet*

SOEKOEL NGANGO
See *hulu babah buya*

SOEKOEL SIMPOEL
See *hulu iku mie*

SOEKOEL TAKA KOEDO
See *hulu tapa guda*

SOEMARA
See *sumara*

SOEMPI
See *sumpi*

SOEROEK
See *suruk*

SONAGANG-KLEWANG
See *ladieng*

SONDANG
See *sundang*

SONGKOK
SULAWESI
A helmet with large protrusions resembling buffalo horns. See also *tandu-tandu*.
(DRAEGER)

SONRI
See *alamang*

SOPOK
KALIMANTAN, DUSUN
The Dusun term for both a spear and a blow-pipe.
(ROTH 1896B; STONE)

SOSOAT
See *soat*

SPEAR
Apart from the spears of which the specific names are known (and as such included in this survey), other types of spears are found all over the archipelago. The shapes and decorations vary greatly. A number of these spears are depicted below.

512. Spear
Sumatra, Acheh. RMV 3600-1058. Acq. in 1959 through the Ethnografisch Museum van de Koninklijke Militaire Academie, Breda. L. 183 cm.

TRADITIONAL WEAPONS OF THE INDONESIAN ARCHIPELAGO

513. Spear
L. 178 cm, blade L. 42.5 cm.

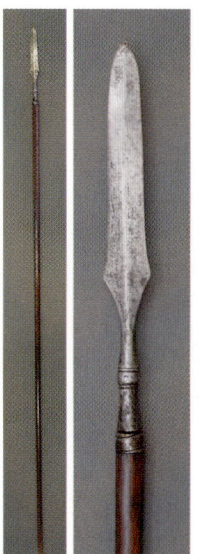

514. Spear
Sumatra, Batak. L. 177 cm, blade L. 27.5 cm.

515. Spear
L. 158 cm, blade L. 30.5 cm.

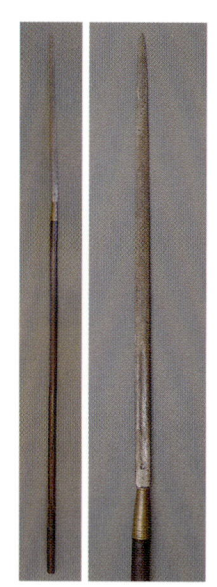

516. Spear
L. 186 cm, blade L. 60 cm.

517. Spear
Kalimantan. L. 216 cm, blade L. 26 cm.

518. Spear
Maluku. L. 192 cm, blade L. 17 cm.

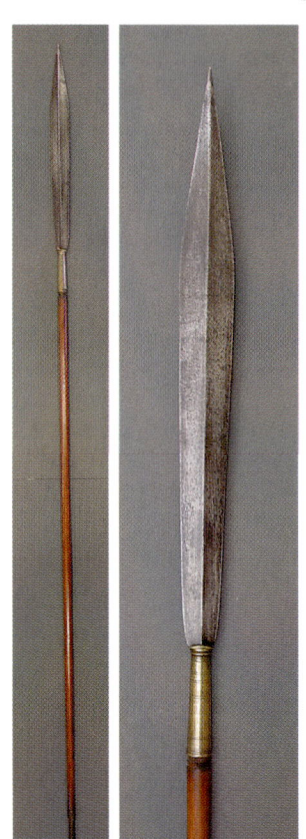

519. Spear
L. 162 cm, blade L. 43.5 cm.

520. Spear
L. 213.5 cm, blade L. 18 cm.

521. Spear
Maluku. Warrior with a spear.

522. Spear
Java. Spear head with a gold encrustation.

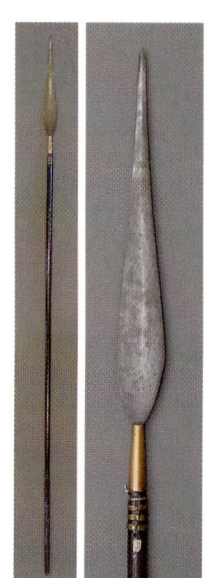

523. Spear
Maluku. L. 201 cm, blade L. 40 cm.

SREPANG
See *sarampang*

STICK SWORD
[TOA, TOPO]

ADORANA, FLORES, SOLOR, SOUTH-EAST SULAWESI

On a number of isles of Nusa Tenggara a type of sword is found characterised by an extremely long, stick shaped, straight hilt. Its length can exceed 1 metre.
On Adonara and Solor the blade somewhat resembles a broad machete. It has a short, heavy back which curves to then bend towards the edge in a long flowing S-shape. The blade is very narrow at the hilt, followed by a shallow indentation to become broader reaching a maximum width of 8-9 cm. The edge is slightly convex.

The grip is a long wooden stem. It is interrupted by, or inlaid with yellow, green or black rings and strips of buffalo horn. Occasionally the hilt has an iron sleeve at the base. Both hands are used when striking. The heavy, broad blade has the effect of an axe and can be lethal.

On Flores, especially in the region around Ende and east thereof, a similar type of sword is found, but here the blade is much larger. It begins narrowly at the hilt and then broadens. Its back is usually straight and somewhat thicker than the rest of the blade. Near the tip the back makes a curved and somewhat concave line towards the edge. Occasionally blades have a fine *pamor*. Almost always these blades are imported from south-east Sulawesi, mostly from the Rumbia region, where swords with blades of this type are called *toa*.

The hilt of the Flores type is a smooth wooden stick, sometimes with an iron sleeve at the base, next to the blade. The hilt is the thickest there to then become thinner towards the tip. Sometimes this tip also has an iron sleeve. The hilt's end is almost always pierced with a hole through which an iron or copper ring is placed. The wood may also have here strips or rings of buffalo horn, or be decorated with punched in, small copper nails. Sometimes a band with suspended goat's or horse's hair is attached to the hilt. A smaller variant of the Flores type has a curved hilt made of buffalo horn and is called *topo*.
(TEXT: K.H. SIRAG)

524. Stick sword (topo)
Flores, Kotta. Warrior with a sword (*topo*), of which just the hilt is visible.

525. Stick sword
Flores. L. 89.5 cm.

526. Stick sword (topo)
Flores. L. 76 cm.

527. Stick sword
Solor. L. 74 cm.

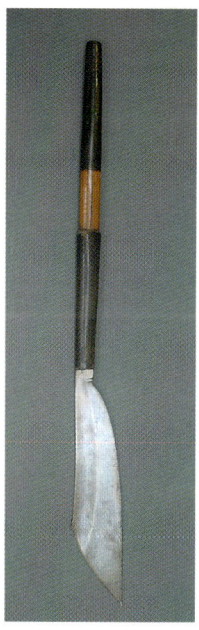

528. Stick sword
Solor. L. 87.5 cm.

SUASA
A alloy of gold, silver and copper used to make hilts (generally filled with resin) or to decorate hilts and scabbards.

SUDAS
See *ranjau*

SUDUK
[SOEDOEK]
A type of knife.
(STONE)

SUHUL
See *sukul*

SUKUL
[SOEHOEL, SOEKOEL, SUHUL]

SUMATRA, BATAK
Suhul means 'hilt' in the language of the Toba Batak. The Karo Batak use the term *sukul*.
(FISCHER 1914)

SUKUL DJENGKAL BENGKOK
See *sukul jengkal bengkok*

SUKUL DJERING
See *sukul jering*

SUKUL DJERRING
See *sukul jering*

SUKUL GERPONG
See *hulu puntung*

SUKUL JENGKAL BENGKOK
[SUKUL DJENGKAL BENGKOK]

SUMATRA, BATAK, TOBA, NORTH-WEST KARO
An S-shaped hilt ending in an oblique, flattened tip found, for example, on the *raut*.
(KREEMER; VOLZ 1909)

529. Sukul jengkal bengkok
Sumatra, Batak. *Raut* hilt.

SUKUL JERING
[SOEKOEL DJERRING, SUKUL DJERING]

SUMATRA, BATAK, KARO, PAKPAK
A hilt which bends at a right angle at the top, ending broadly and somewhat bluntly. This hilt is sometimes in the form of a stylised bird's head. It is found, for instance, on the *kalasan situkas* (Karo), the *amanaremu* (Karo, Pakpak), the *candong* (Pakpak), the *ladingin* (Pakpak) and the *andar andar*.
(KREEMER; VOLZ 1909)

530. **Sukul jering** Sumatra, east Karo. *Kalasan situkas* hilt.

531. **Sukul jering** Sumatra, Karo Batak. *Kalasan* hilt.

532. **Sukul jering** Sumatra, Pakpak Batak. *Julung julung* hilt.

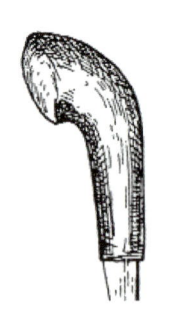

533. **Sukul jering** Sumatra, south west Karo. *Amanremu* hilt.

SUKUL KATUNGANGAN
SUMATRA, BATAK, PAKPAK
A thick horn hilt, curved in a right angle. Immediately after the bend, the upper part splits into two opened lips in the shape of a dragon's mouth. In the middle sometimes a short semi-circular knob occurs. At the base the hilt has a broad metal ring. These hilts are found, for instance, on the *ladingin* (Pakpak) and the *katungung* (Pakpak).
(VOLZ 1909)

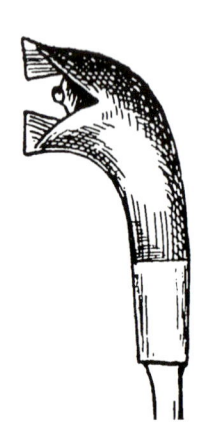

534. **Sukul katungangan** Sumatra, Pakpak Batak. *Katungung* hilt.

535. **Sukul katungangan** Sumatra, Pakpak Batak. *Ladingin* hilt.

SUKUL LAPAN SAGI
See *hulu lapan sagu*

SUKUL MEKEPIT
See *hulu meu apet*

SUKUL NGANGA
SUMATRA, BATAK, KARO
A hilt with a slight curve in the centre. The end is formed by two protrusions which run parallel on the inside and apart from each other on the outside. The protrusions are flattened on the sides. This type of hilt is found with the *kalasan* and the *andar andar*.
(KREEMER; VOLZ 1909)

536. **Sukul nganga** Sumatra, Batak. *Andar andar* hilt. L. 13 cm.

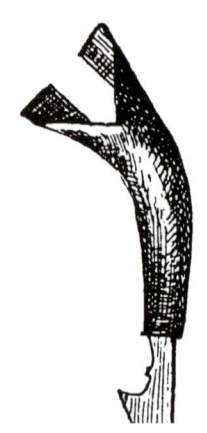

537. **Sukul nganga** Sumatra, Karo Batak. *Kalasan* hilt.

SUKUL NGANGAN
SUMATRA, BATAK, KARO, PAKPAK
A thick hilt with a slight curve in the middle. The end consists of two almost parallel sturdy protrusions forming a mouth of sorts. In the middle, near the 'mouth's' base, there is sometimes a bulbous knob. This hilt is found, for instance, with the *ladingin* (Pakpak), the *corik* (Pakpak), the *andar andar* (Karo), the *kalasan* (Karo) and the *rudus* (Karo).
(VOLZ 1909; KREEMER)

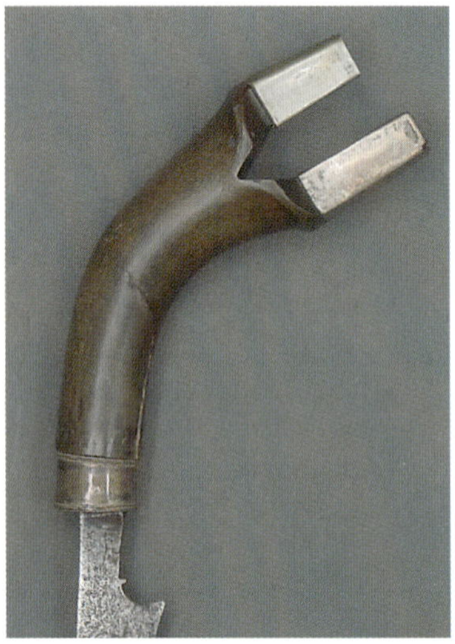

538. **Sukul ngangan** Sumatra, Batak. *Andar andar* hilt. L. 12 cm.

SUKUL NGANGO
See *hulu babah buya*

SUKUL SIMPOEL
See *hulu iku mie*

SUKUL SIMPUL
See *hulu iku mie*

SUKUL TANKE DJAMBE
See *sukul tanke jambe*

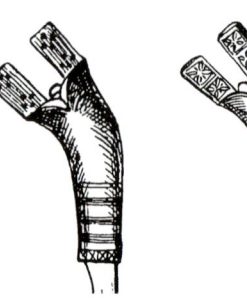

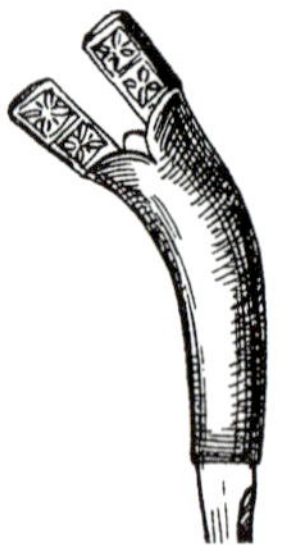

539. **Sukul ngangan** Sumatra, Pakpak Batak. *Ladingin* hilt.

540. **Sukul ngangan** Sumatra, Simsim, Batak. *Rudus* hilt.

SUKUL TANKE JAMBE
[SUKUL TANKE DJAMBE, TANKE DJAMBE]

SUMATRA, PAKPAK BATAK, KARO HIGHLANDS

A type of hilt occurring with, for instance, the *andar andar* in north-west Karo. The form is plain, round and slightly curved, but broadens towards the end. The end is somewhat flattened.
(KREEMER; PARAVICINI; VOLZ 1909)

542. **Sukul tanke jambe**
Sumatra, north west Karo. *Julung julung* hilt.

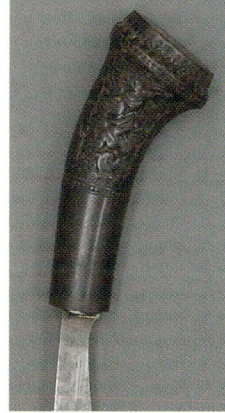

543. **Sukul tanke jambe**
Sumatra, Batak. *Rawit* hilt. L. 9.5 cm.

SUKUL TANKE TERB
SUMATRA, BATAK
A type of hilt originating from Taneh Kembaren and Merdinding.
(VOLZ 1909)

SUKUL TAPA KOEDO
See *hulu tapa guda*

SUKUL TAPA KUDO
See *hulu tapa guda*

SULU KERIS
See *sundang*

544. **Sumara**
Sulawesi. L. 59.5 cm.

SULU KLEWANG
A sword with one sharp side which broadens and becomes heavier towards the point. The edge is straight. The blade's oblique end often has a pointed protrusion. The hilt is made in such a way that the sword can be used either with one or two hands.
(GARDNER 1936)

SULU KNIFE
See *barong*

SUMARA
[SIMALA, SOEMARA]

SULAWESI, BOLAANG MONGONDO, EAST FLORES

A sword related to the *penai*. Its blade has a straight edge and back. Both run apart towards the point so that the point is a little broader than the base. Its back runs to the edge in an angle. The form of such blades is known as *kajeli*. The hilt makes a 90° bend and ends in two long protrusions standing wide apart, creating the shape of an open mouth. The scabbard (*guma*) is straight, with a small slightly asymmetric foot.
(FOY; JUYNBOLL 1927)

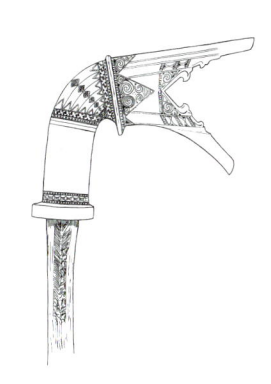

545. **Sumara hilt**
Sulawesi.

SUMPI
[SOEMPI]

SULAWESI, TORAJA

A bamboo blow-pipe with a length of 30-40 cm. The dart is made of a splinter of wood. On its end we see a cone made of fibre of the *banga* palm. Its tip is covered with poison (*ipoh*). The range of this blow-pipe is c.25 metres. All Toraja use blow-pipes as combat weapons.
(LOEBÈR 1928; DRAEGER)

SUMPING
MALUKU, BURU
A blow-pipe for poisonous darts, perhaps a combat weapon, but more likely used for hunting.
(DRAEGER)

SUMPIT
See *sumpitan*

SUMPITAN
[LEPUT, LOHINGLAMBI, PUT, SIPET, SUMPIT]

KALIMANTAN

The *sumpitan* is a blow-pipe made of a single piece of hardwood, preferably of the *niagang*-tree which has a straight grain and hardly any knots. A rough stick, measuring c.150-240 cm and c.7-10 cm in diameter, is cut from a tree trunk. It is processed on the spot or taken home to be further worked on. In the first case, a structure is built on which the maker of the hole is seated. This stick is secured vertically in such a way that the upper part only just protrudes from the platform. The lower part rests on the ground. Next, a long straight iron rod is used to bore and chisel a hole from above. Sometimes, however, the hole seems to be bored from below. In order to keep the rod in a vertical position, means of support with various heights are made from forked branches. Loose wood chips are removed from the bore hole using water. In order to bore the last part of the hole the stick is slightly curved. The result is that the hole is not completely straight once the work

546. Sumpitan
Kalimantan, Sarawak, Sea Dayak.

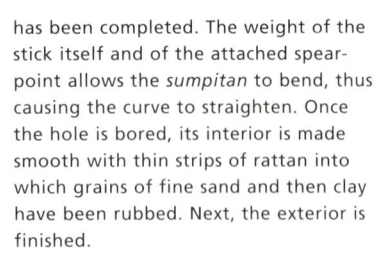

548. Sumpitan
Kalimantan. L. 226 cm, blade L. 23.5 cm.

has been completed. The weight of the stick itself and of the attached spear-point allows the *sumpitan* to bend, thus causing the curve to straighten. Once the hole is bored, its interior is made smooth with thin strips of rattan into which grains of fine sand and then clay have been rubbed. Next, the exterior is finished.

To the tip a spear-point is attached which is tied to the exterior using rattan. This tip is almost always made of iron, but sometimes a tip made of ironwood (*Eusideroxylon zwageri*, Fam. *Lauraceae*) occurs. In the binding, opposite the spear tip, we often find a wooden, bone or metal 'foresight' (*klahulon*) in order to aim better. Using resin, a *cowrie* shell is sometimes attached to the shaft serving as a foresight. The mouth-piece may have a horn ring or a ring of resin (*damar*). The *sumpitan* is used to fire darts at birds and small game, but also in combat.

The *sumpitan* comes with blow-darts (*langa*), a quiver (*tolor* or *tavang*) and a small gourd (*hung*) in which are stored the small cones of the pith of a tree or of a certain species of thorny creepers which are attached to the back of the darts. These serve to close the bored holes so that the darts can be blown out

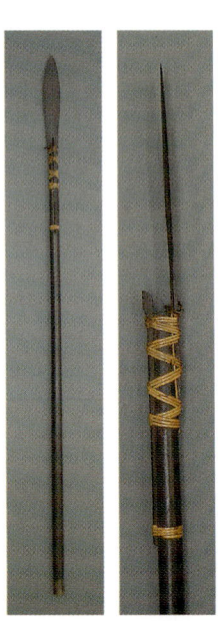

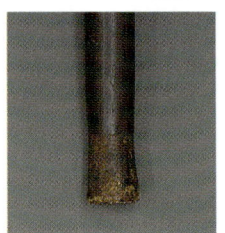

547. Sumpitan
Kalimantan. *Sumpitan*; side view of the blade; mouthpiece of *damar*.
L. 138.5 cm, blade L. 24.5 cm.

549. Sumpitan
Kalimantan. L. 193 cm, blade L. 27.5 cm.

forcefully. Furthermore, instruments are used to prepare the poison (*ipoh*) and to apply it to the tips of the darts. There we find a small instrument to mould the cones into the correct size making them fit exactly into the bored hole of the *sumpitan*.

The darts have a range of c.35-54 m, but for a direct hit the range is c.20-25 m. When leaving the blow-pipe, the darts allegedly have a speed of 180 km per hour. The *sumpitan* is also known as *sumpit* (Iban), *leput* (Kayan), *sipet* (Ngaju) and *put* (Punan).

(AVÉ; DRAEGER; FURNESS; GARDNER 1936; GOMES; HEIN 1890; HOSE 1912; KEPPEL 1846A,B; LOW; ROTH 1896A,B; SKEAT; SOPHER; STONE; VOLZ 1909, 1912; WEIGLEIN)

SUNDANG
[SONDANG, SULU KERIS]

KALIMANTAN, SULAWESI, PHILIPPINES, SULU ARCHIPELAGO

A heavy type of *keris* to stab or cut with. The blade has a round tang. In order to prevent the blade from turning in the hilt when used as a cutting weapon, one or two strips (*sigi*) are attached to hold the blade firmly to the hilt. This connection is characteristic of the *sundang*.

The blade has two sharp sides. It may be either waved or straight. It can be smooth, or have ribs or channels lengthwise. Its point is blunt and its base is asymmetrically broad. The least protruding side has a stylised elephant trunk, the other side has a notched rim. The hilt has a straight, round lower part often wound around with thread or metal wire. Its knob usually has the shape of a stylised bird's head. The scabbard has a broadened upper part into which the blade's base fits.

(CATO; DIELES; ENGEL; FORMAN; FOY; GARDNER; STONE)

550. Sundang
Kalimantan (?). Hilt with a kakatua-pommel. L. 67.5 cm.

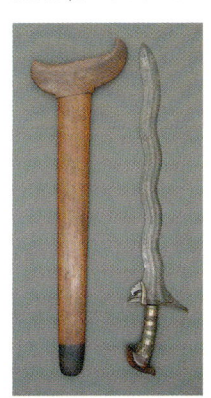

551. Sundang
Kalimantan (?). Hilt with a kakatua-pommel. L. 57.5 cm.

552. Sundang
Kalimantan.

SUNI
TIMOR
A type of sword.
(FOY)

SURIK I
SUMATRA, BATAK

A sword with a straight backed blade and a somewhat S-shaped curved edge. The blade broadens towards the point and curves towards the back at the point. The hilt is slightly curved halfway to end in two parallel short protrusions. Its bottom half has a tapering metal sleeve which broadens out near the blade. The scabbard is broad and flat. Lengthwise an elevated rib runs. At the point the scabbard curves somewhat backwards. Its mouth broadens slightly on the side of the edge. See also *piso sanalenggam*.

(ROGERS; VAN DER TUUK)

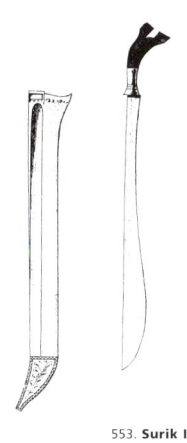

553. Surik I
Sumatra, Batak.

SURIK II
CENTRAL TIMOR

A sword with a curved blade which, including the scabbard, can reach a length of more than 1 metre. *Suriks* were worn only by the Meos, the foremost fighters, and usually also the most successful head-hunters of the villages.

The blades are often of Dutch, English or Portugese origin, but

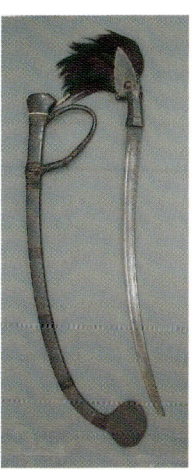

554. Surik II
Timor. L. 93 cm.

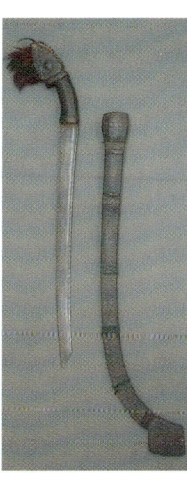

555. Surik II
Timor. L. 73.5 cm.

556. Surik II
Timor. L. 79 cm.

reduced in length. In the Dutch blades one may find the VOC emblem and a date.

The hilt has the shape of a high triangle above a hexagonal or octagonal shaft which is broader at the foot. In the triangle's upper sides holes are bored into which tufts of black or red dyed goat's or horse's hair are fixed. In the middle of the flat sides of this triangle we often find a raised circle: the 'eye' of a mythical animal.

The scabbard follows the blade's line and is made of two long pieces of wood held together by rattan ties. The open parts between these ties are often decorated with geometrical carved lines. Its mouth is always much thicker and more convex than the rest of the scabbard. This mouth has a somewhat elevated rim along the top into which the hilt's lower part exactly fits. It also has carved decorations. The scabbard is much longer than the blade and ends in a broad, flat foot. The foot is round, rectangular, or a combination thereof, and often decorated with carvings in the same patterns as the rest of the scabbard. Sometimes holes are bored in the rims of the foot in which, as with the hilt, tufts of hair are fixed. See also the entry: swords of the Timor group.

(TEXT: K.H. SIRAG)

SURUK
[SOEROEK]

TANIMBAR ISLES
An ancient type of sword, used during marriage ceremonies.
(LEENDERTZ; RODGERS)

SWORD
A large number of swords without a specific name are found all over the archipelago. A selection hereof is depicted below.

557. Sword
West Java. L. 66.5 cm.

558. Sword
West Java. L. 67 cm.

559. Sword
Sumatra, Batak. L. 71 cm.

560. Sword
Maluku (?). L. 44 cm.

561. Sword
West Java. L. 65.5 cm.

562. Sword
West Java. L. 56.5 cm.

563. Sword
Sulawesi. L. 62 cm.

SWORD

SUMBAWA, BIMA

Swords originating from Sumbawa have blades which broaden towards the point. The back runs towards the edge near the point. The scabbard broadens both at the mouth and the foot.
(FISCHER AND RASSERS 1924)

564. Sword
Sumbawa, Bima. L. 52.5 cm.

565. Sword
Sumbawa, Bima. L. 51.5 cm.

SWORDS OF TANIMBAR

On Tanimbar no real forging culture exists. Small local smithies are found on various islands. However, only simple products are made there such as: fish spears, points for arrows and lances, agricultural tools and machetes.

Many of the more beautiful sword blades are imported from south-east Sulawesi and Maluku, or are made by travelling smiths from Seram. They sail to the various islands and produce utensils (including sword blades) in make-shift smithies on the beach. The iron used (mainly scrap from ships) has been collected by the inhabitants of Tanimbar themselves. Due to the variety of origin of the smiths, the blades on Tanimbar vary greatly in shape.

Most blades found on Tanimbar have many similarities with blades from Maluku and south-east Sulawesi. These blades are slender at the hilt and broaden towards the point. Their backs bend near the point in a slight curve towards the edge, or make a sharp turn to then run in a straight line towards the edge. These backs are always straight, the edges are often somewhat convexly curved. The blade's length is mostly c.50 cm, but 65 cm long blades are also found.

Less frequent are blades with similarities with those from Seram, which are narrower and slightly curved at the tip. The edge of this form is somewhat convex, the blade broadens towards the tip. The back is straight at first. Next, at c.12 cm from the tip, runs in a concave line towards the edge, so that a crescent shape is formed ending in a sharp point. Moreover, blades of European origin are

566. Sword of Tanimbar
Tanimbar. RMV 1249-12. Acq. in 1900. L. 81 cm.

used as well as blades of machetes traded by Chinese.
Hilts from Tanimbar are nearly always made of wood, but sometimes horn is used. They can be divided in two main types:
(a) with a straight shaft, somewhat thickened at the base or with a metal sleeve. At the top the hilt curves towards the edge and runs into a very stylised monster's head with a V-shaped open mouth. The upper lip is usually longer than the lower lip. The mouth is often filled with a small round tongue, pierced or not. Sometimes it is decorated with pieces of string or red cotton thread. Now and then simple geometrical carving is applied or an inlaid bead serves as 'eye'.
(b) also with a straight shaft, curving at the top towards the edge. The monster's mouth is closed here and thus takes the form of a flat butt-end of a pistol. The upper part has carved grooves running along it. Sometimes it is inlaid with small triangular bits of mother-of-pearl. Pieces of rope or small patches of red cotton are often attached to the top.
In very rare cases an old sword has a hand protector made of tanned sting-ray or shark's skin to which shells are attached as decoration.
Imported hilts, often with matching blades, from Sulawesi or even Timor are also found. Their scabbards are made on Tanimbar at a later stage.
A Tanimbar scabbard is made of two parts of wood, tied together by means of (woven) rattan strips or rope. Its upper part is often wound around with metres long strips of cotton. On its top part one nearly always finds a small wooden elevation in which two holes for a carrying rope are bored. This elevation is often straight, sometimes carved in an angled S-shape. At the top, the scabbard is slightly broader and protrudes somewhat towards the edge. This protruding part is sometimes slightly curled upwards. The scabbard's lower part can have three different forms:
(a) straight or cut off somewhat obtusely at the foot;
(b) broadening a little to the end, and with at its sides in the centre, a slightly elevated rib lengthwise;
(c) becoming thinner towards the end and bending a little towards the blade's back. It ends in a sharp or round point. These

567. Sword of Tanimbar
Maluku, Tanimbar. Warrior carrying a sword.

568. Sword
Tanimbar. L. 65 cm.

569. Sword
Tanimbar. L. 63.5 cm.

570. Sword
Tanimbar. L. 58 cm.

571. Sword
Tanimbar. L. 65 cm.

572. Sword
Tanimbar. L. 73.5 cm.

573. Sword
Tanimbar. L. 68 cm.

574. Sword
Tanimbar. L. 63 cm.

575. Sword
Tanimbar. L. 53 cm.

The blade may be straight or curved, like a sabre. Often sabres of English, Portugese or Dutch origin were used. With blades produced locally the back almost always runs towards the edge. Scabbards of type (a) are, as are the blades, straight or curved and made of two wooden parts, held together by rattan bindings. Its lower part ends in a broad, flat foot protruding both on the edge and on the rear side. This foot often has carving work. Its mouth is always much thicker and rounder, with on the top plane a carved out deeper part into which the hilt's lower part fits exactly. This mouth is nearly always carved with S-shapes, diamonds, and globules. An extremely long and curved variant of this type is called *surik*. The matching scabbard's lower part ends in a large, flat, round or somewhat angled foot;

579. Sword hilt of the Timor group, type a
Timor. L. 19.5 cm.

scabbards often have blades with a crescent-shaped end. Only geometrical lines, triangles or small circles occur near the scabbard's mouth. Although excellent carvers are found on Tanimbar, carvings on scabbards hardly ever occur.
(TEXT: K.H. SIRAG)

SWORDS OF THE TIMOR GROUP
ALOR, PANTAR, ROTI, SAVU, TIMOR, WETAR

Five main types of swords are found on the isles of the Timor group, mainly recognisable by the shape of their hilts:

(a) as is found everywhere on Timor. The hilt is made of buffalo horn and has a straight shaft. Its top is shaped as a triangle, protruding towards the edge. Its lower part broadens all around. In the centre of the triangle an elevated 'eye' occurs, which may also have the shape of a rosette or flower. The flattened sides often have bored holes in which tufts of goat's hair are fixed. The middle, top and bottom of the shaft are decorated with geometrical carving (S-shapes, loops and cables). A rare variant is the one in which the upper side is slightly curved, and pierced underneath so that it resembles the banisters of a staircase.

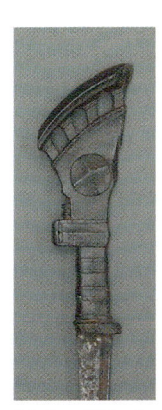

576. Sword hilt of the Timor group, type a
Timor. L. 23 cm.

577. Sword hilt of the Timor group, type a
Timor. L. 18.5 cm.

578. Sword hilt of the Timor group, type a
Timor. L. 16 cm.

580. Sword of the Timor group, type a
Timor. L. 74 cm.

581. Sword of the Timor group, type a
Timor. L. 70 cm.

(b) as found on the isles of Pantar, Alor and Wetar. It is a variant of (a) as its triangular part is somewhat more tilted than of (a). The upper part there, is the edge with (a). On Alor and Wetar the triangular part's edge runs almost parallel with the blade. On Pantar this side curves towards the blade's edge. The Pantar type is much longer than the one from Alor and Wetar, while the point at the hilt's top is much blunter. The central 'eye' on Pantar often has a flower or rosette shape. On Alor it is mostly only a circle, while on Wetar a diamond-shape also occurs. On (b) tufts of hair are also found, but far fewer than on (a). The

582. Sword of the Timor group, type b
Pantar. L. 64 cm.

583. Sword of the Timor group, type b
Pantar. L. 62 cm.

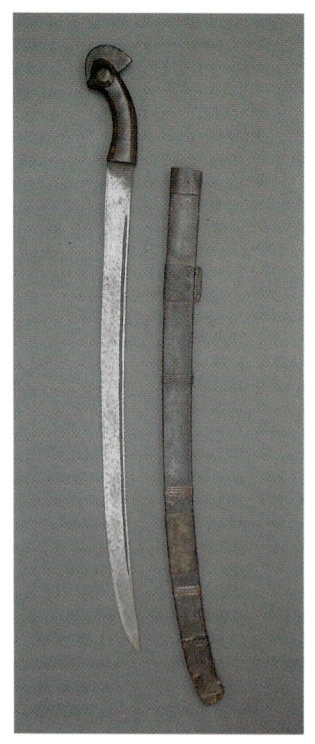

585. Sword of the Timor group, type c
Timor. L. 75.5 cm.

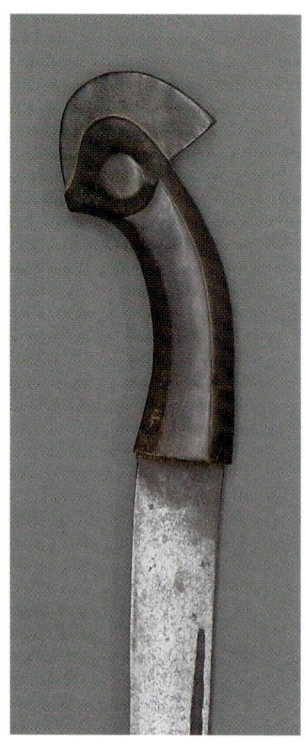

586. Sword hilt of the Timor group, type c
Timor. L. 15 cm.

carving work is less fanciful. The blades are produced locally and are almost always straight. They can be equally straight over the entire length, but may also broaden somewhat towards the point. The back curves towards the edge.

The scabbard is made of two wooden parts, held together with rattan. Its mouth has a protrusion facing towards the edge. On Alor and Wetar this is rectangular, on Pantar this protrusion often ends in carving in the form of feathers and curls. The sword with this type of hilt found on Alor and Wetar is also called *rugi*;

(c) as found on Central Timor. It has the shape of a bird's or cock's head, mostly very stylised. Slightly curved towards the edge, it ends in a round bird's head. The blunt beak forms the most protruding part. At the top we see a kind of cock's comb. In the middle of the head an elevated circle is carved which may be so large it covers the entire head. Sometimes the 'eye' consists of a silver coin.
The blades are often of English, Portugese or Dutch origin. Straight blades are not common.
The scabbards are made of two parts of wood, held together by (usually) silver-coloured metal strips. Nowadays aluminium is

584. Sword of the Timor group, type b
Alor. L. 68 cm.

used. The scabbard does not have a broadened foot. Its mouth is scarcely broader than the rest of the scabbard, and is just as broad as the hilt's lower part;

(d) as found on south-west Timor, with Camplong and Niki-Niki as a centre. It depicts a stylised animal's head. Its shaft is mostly round or oval in cross-section and at the bottom somewhat thicker. It is often provided with a metal strip or cuff. At the top the hilt curves towards the edge. This curve may vary from slight to an angle of almost 90°. The end is blunt and round, with a groove in it to represent a mouth.
The most striking element of this hilt is a shell-shaped protrusion on top of the back, where the curve begins. This protrusion has become a blunt point. With very stylised examples the mouth is lacking. The blades are often short to very short. Broken off and re-ground sabres are often used. Straight blades are less frequent than curved ones.
The scabbards are made of two wooden parts, held together by broad strips of silver-coloured metal or brass. Its lower part is flat. Its mouth has at the edge side a triangular broadening, sometimes with a V-shaped incision at the rim. This type of weapon

587. Sword of the Timor group, type d
Timor, Niki Niki. L. 53.5 cm.

588. Sword hilt of the Timor group, type d
Timor. L. 13 cm.

back somewhat less broadened, or hardly broader than the diameter of the hilt's stem. This mouth is always decorated with carving resembling that on the hilt, also with a raised central 'eye'. The scabbard has an angled foot, facing towards the edge side. This foot is decorated with carving which always includes a central raised circle.

On Savu this type of sword is called *hemola*, on Roti it is called *tafa*.

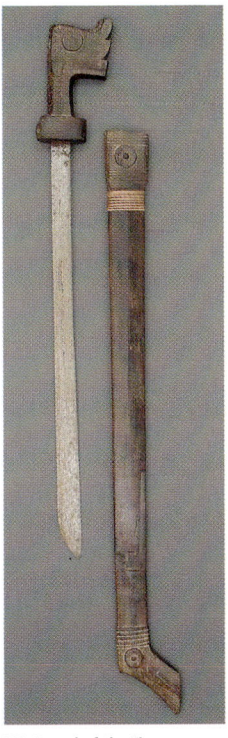

589. Sword of the Timor group, type d
Timor. L. 43.5 cm.

often has a belt fixed under the broadened part of the scabbard. The belt runs across the shoulder, so that the weapon is hung under the armpit;

(e) as found on the isles of Savu and Roti. As are all hilts of the Timor group, type (e) is made of buffalo horn, but it is here yellowish instead of black. It has a surprisingly short shaft, the lower part of which is quite broad. At the top we find a large, flat almost square part protruding towards the edge and resembling a kind of flag. The edge's side is incised with two shallow, oblique V-shaped indentations. The square's sides are, with the exception of the lower side, pierced and provided with tufts of goat's hair. The entire hilt is decorated with carvings. In the centre of the sides we see an elevated circle with a protruding central point.
The blades are of local provenance and always straight. The back curves towards the edge.
The scabbard is made of two wooden parts, tied together with rattan. The binding can continue over the scabbard's entire length, but can also be interrupted by carving. Its mouth can be very broad, and towards the

590. Sword of the Timor group, type e
Roti or Savu. L. 60 cm.

591. Sword of the Timor group, type e
Savu. L. 73 cm.

On Timor swords are also found with a very stylised pistol butt-end shaped hilt, without any carved decoration. Their scabbards are not carved and are only decorated with silver-coloured metal strips. Curved, sabre type blades are seemingly preferred.
Hilts and scabbards of the swords of the Timor group were made by professional carvers and display a high degree of skill.
(TEXT: K.H. SIRAG)

592. **Tandu tandu**
Sulawesi, Poso area. RMV 43-9. Acq. in 1864 from C.B.H. von Rosenberg. H. 10 cm.

T t

ALPHABETICAL SURVEY

TA MING
SULAWESI, MINDANAO (PHILIPPINES), SUBUNAN
A round shield.
(STONE)

TABAK
MALUKU, HARUKU
A small stick, sharpened on both points and hardened in a fire. The *tabak* is used as a sword or as a *ranjau*.
(DRAEGER)

TACULA TEFAO
See *takula tofao*

TADJI
See *taji*

TAFA
ROTI
A sword the scabbard of which has a foot protruding to one side. This distinguishes the *tafa* from the swords of, for example, west Timor. See also the entry: swords of the Timor group.
(FOY)

TAJI
[TADJI]

BALI
A stiletto with a slender blade. Sometimes poison is applied to its point.
(DRAEGER)

TAKA BLADE
SUMATRA, KARO BATAK
A type of blade with a straight back and an S-shaped edge with a sharp point. Near the hilt, the edge turns in towards the back and curves into a small dip, giving the blade a narrow base.
(VOLZ 1909)

593. **Taka blade**
Sumatra, Karo Batak.

TAKOELA SINALI
See *takula sinali*

TAKULA SINALI
[TAKOELA SINALI]

CENTRAL NIAS
A war hat made of very tightly woven *gnetum* fibres. The shape is cylindrical with a flat top and a sharp rim. On the lower side three rectangular protrusions of the same material are attached, two above the ears, and a larger one to protect the neck.
(FISCHER 1920)

TAKULA TOFAO
[TACULA TEFAO]

CENTRAL NIAS
A steel bowl-shaped helmet, usually made of pieces of sheet iron linked together, sometimes by metal wire. It has a narrow rim and sometimes one or more

594. **Takula tofao**
Nias. Warrior carrying a *takula tofao* (helmet), a *baluse* (shield) and a *toho* (spear).

595. **Takula tofao**
Nias. RMV 1002 123. Acq. in 1894 from Palmer van den Broek. H. 13.5 cm.

dentated rows of combs. It may have ear protectors. Often the *takula tofao* is decorated with huge iron ornaments in the shape of branched plumes, in the form of a tree of life, or other fanciful figures or protrusions.
(FELDMAN 1990; STONE)

TALAWANG
KALIMANTAN
A shield in the shape of a *kliau*, the decoration of which represents a demon's head.
(HEIN 1890; STONE)

TALEMPAK
A short lance carried by a section of female soldiers belonging to the Amazon army of the queen of Wana Siluman in Galuh.
(HAZEU)

TALI PEDANG
A belt for a sword, carried around the waist, so that the sword hangs on the left side.
(GARDNER 1936)

TAMBENG
SULAWESI, TORAJA
A round, wooden shield with protruding metal spikes on the front. It may also be made of leather or rattan, woven on a wooden frame. This shield is often lavishly decorated.
(DRAEGER)

TAMBUK
SULAWESI, TORAJA
An oval shield with narrow tapering points. It is made of wood, leather or of plaited rattan on a wooden frame.
(DRAEGER)

TAMBULOH
KALIMANTAN
A spear with a small point on a long iron shaft.
(HEIN 1890)

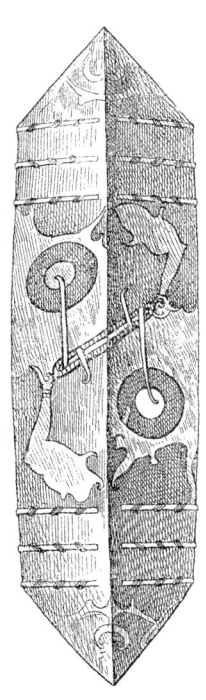

596. **Talawang**
South east Kalimantan.

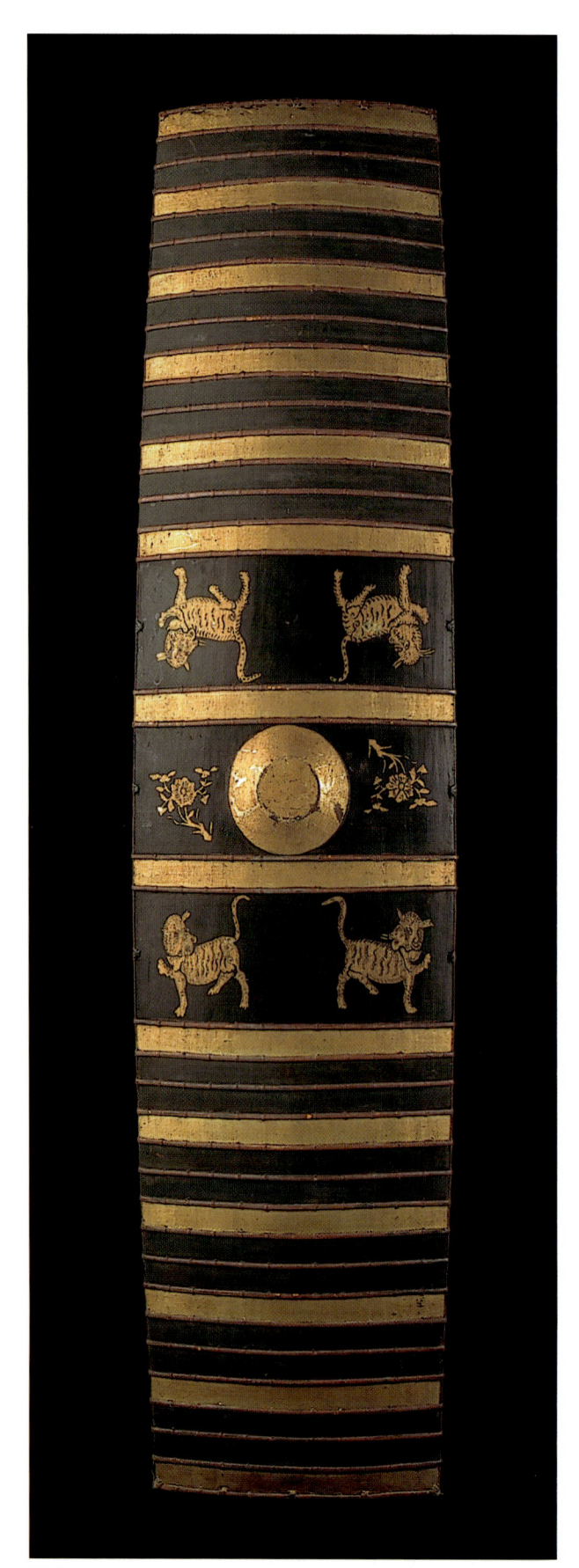

TAMENG

JAVA, LOMBOK, TIMOR

A shield with a round or elongated shape. The long shape is an old type (before c.1800). Its sides are slightly convex. Both ends are almost straight. In the middle of the back, over the entire length, a reinforcement is placed in which the grip is situated. The front has a knob in the centre.
The round form is a later model and may be made of rattan covered with leather or cloth. Often it is provided with small metal stars or spikes. Sometimes the *tameng* is made of brass.
(DRAEGER; GARDNER 1936; RAFFLES; STONE)

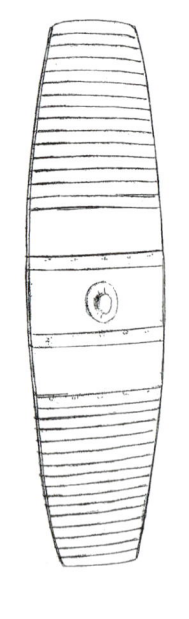

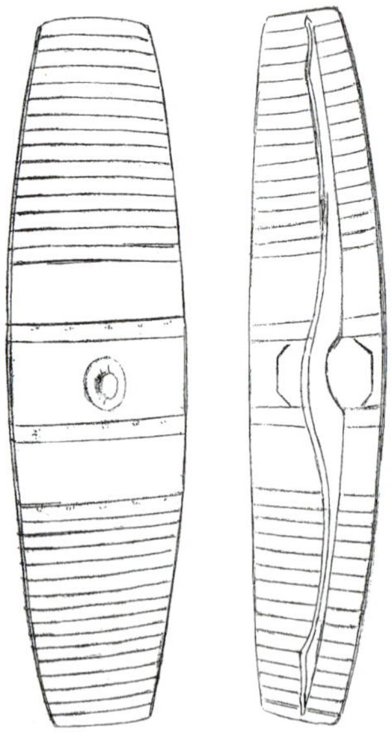

TAMPELAN

NORTH SULAWESI, BOLAANG MONGONDOW

A sword with a blade of which the edge and back are straight, broadening somewhat towards the point so that the centre of gravity lies at the end. The back runs obliquely towards the edge. Halfway along this oblique plane a protrusion is found. The hilt has a large flat decoration, sometimes with carvings and tassels of black and brown horse's hair. The hilt may or may not have a crossbar as its base. See also *kampilan*.
(FOY)

TAMUA

SUMBA, SUMBAWA

A round shield made of buffalo hide. Its rim is edged with rattan.
(STONE)

598. **Tameng**
Java.

597. **Tameng**
Java. RMV 454-20. Acq. in 1884 from K. Hoogeveen. H. 140 cm.

599. **Tamua**
Sumba / Sumbawa. Diam. 52 cm.

600. **Tamua**
East Sumba. RMV 858-150. Acq. in 1891 through the Koninklijk Nederlandsch Aardrijkskundig Genootschap, Amsterdam. Collected 1890-1891 by dr. H.F.L. ten Kate. Diam. 66.5 cm.

601. **Tamua**
Sumba / Sumbawa. Diam. 66.5 cm.

602. **Tamua**
Rear view of illus. 601.

TANDOE
See *lungkee*

TANDU
See *lungkee*

TANDU TANDU
[ULU-ULU]

SULAWESI, TORAJA, TOBELA

A war hat made of plaited rattan and sometimes covered with the skin of the *tokata* or *kuse* (sloth). It is usually decorated with an erect plume and one or more pairs of metal or thick leather protrusions in the shape of buffalo horns which symbolise the fighter's great power. These head-dresses used during a time of continuous strife, were already a rarity at the beginning of the 20th century. See also *songkok*.
(GRUBAUER; ZAAL)

603. **Tandu tandu**
Sulawesi, Poso area. RMV 43-9. Acq. in 1864 from C.B.H. von Rosenberg. H. 10 cm.

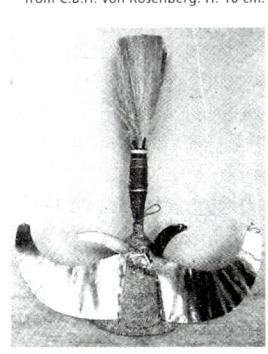

604. **Tandu tandu**
Sulawesi, Matana, Tobela.

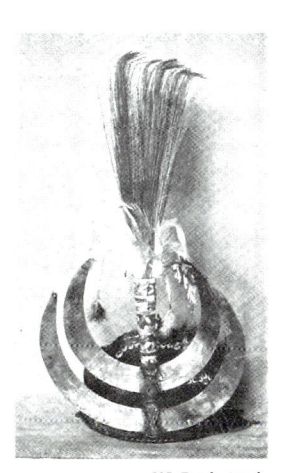

605. **Tandu tandu**
Sulawesi, Matana, Tobela.

TANGIRRI
KALIMANTAN

A blow-pipe with a triangular point.
(STONE)

TANKE DJAMBE
See *sukul tanke jambe*

TAO
SULAWESI, UJUNG PANDANG

A sword with a long, heavy blade sharpened on one side.
(DRAEGER)

TAPAK KUDAK
SUMATERA, ACEH

A sword with a straight-backed blade. Its edge is straight near the hilt and then becomes convex.
(DRAEGER)

TARABADJOI
See *tarah baju*

TARAH BADJOE
See *tarah baju*

TARAH BAJU
[KLEWANG TARAK BAGOE, ROEDOES TARAH BADJI, RUDUS TARAH BAJI, TARABADJOI, TARAH BADJOE]

SUMATRA, ACEH

A type of *klewang* with a slightly concave back and a straight edge. This edge is sharpened along almost the entire length. Its back curves in towards the edge. Its blade broadens towards the point. The *tarah baju* is carried without a scabbard. The Gayo term for this *klewang* is *rudus tarah baji* (meaning: 'cutting of wedges').
(DIELES; GARDNER 1936; JACOBS; KREEMER; VAN LANGEN; NIEUWENHUIZEN; SNOUCK HURGRONJE 1904; VERMEIREN 1983)

TARBIL
A cross-bow with which to fire pellets.
(STONE)

TAVANG
KALIMANTAN, AOHENG

A bamboo quiver for blow-pipe darts with a long hook in order to attach it to the belt. The hook's upper part may have beautiful carving. This hook is fastened to the quiver with bindings. The lid (*tiong bu'u*) is also of bamboo. See also *tolor*.
(SELLATO)

TAWARA
SULAWESI, BOMBA, BADA, KULAWI

A lance with a variety of shapes and sizes. Beautiful examples have shafts made of red hardwood and carvings over the entire length. Lances from Bada and Kulawi are decorated with tassels of goat's hair dyed black-white, black-red or white-red. This hair covers the lance from the middle down to the foot, which has a long iron spike. It is wound into a long rope measuring up to c.5 m, and tied in a tight spiral around the shaft so that the hair stands outwards. For normal use of the lance, the part with hair is kept in a covering of *fuja* or cotton in the same colours as the goat's hair decoration.
(GRUBAUER)

TAWOK
JAVA

A spear with a diamond-shaped point.
(STONE)

TEBOEN
See *anggang totok*

TEBUN
See *anggang totok*

TEKKEN
JAVA, MADURA, BALI

A weapon in the form of a walking-stick.
(DRAEGER)

TEKPI
An originally Chinese weapon made of iron with a large hand protector and a blunt point. The *tekpi* may be used to fend off blows or to strike or beat one's opponent. With this weapon it was possible to avoid the prohibition against carrying edged weapons.
(GARDNER)

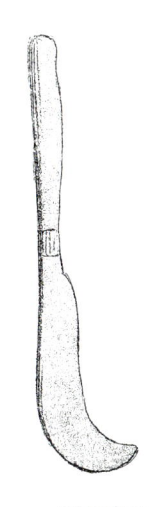

TELABUNA
JAVA

A short, heavy machete with a long round hilt mainly used as an agricultural tool. The blade's back is convex and the edge concave. Its curved point protrudes strongly.
(RAFFLES)

TELAGOE
See *balato*

TELEMPANG
JAVA

A spear with a broad tip. On both sides we see a bent back curve with an iron point on the lower part.
(STONE)

606. **Tawara**
Sulawesi. L. 186.5 cm, blade L. 27.5 cm.

607. **Telabuna**
Java.

TELENGA
See *tolor*

TELEP
[TELEP ANAK PANAH]
A quiver for blow-pipe darts. This quiver has a single or double version.
(SELLATO, STONE)

TELEP ANAK PANAH
See *telep*

TEMBILAH
[TEMBILAN]
KALIMANTAN
A quiver for blow-pipe darts.
(SELLATO, STONE)

TEMBILAN
See *tembilah*

TEMBOELONG
See *tembulong*

TEMBOK LADA
See *tumbok lada*

TEMBONG
A staff or large club.
(GARDNER 1936)

TEMBONG KEMBOJA
A short club.
(GARDNER 1936)

TEMBULONG
[TEMBOELONG]
SUMATRA, ACEH
A type of lance.
(KRUIJT)

TENGKOELOE
See *ketupung*

TENGKULU
See *ketupung*

TEPUS
KALIMANTAN
A blow-pipe with a smooth, sharp point.
(STONE)

TERISULA
See *trisula*

TETE NAULU
NIAS
A war cap made of woven rope of the *aren* palm. It is semi-circular with a flattened top. There is a protruding rim all around. It is decorated with tassels of rope or other materials.
(FELDMAN)

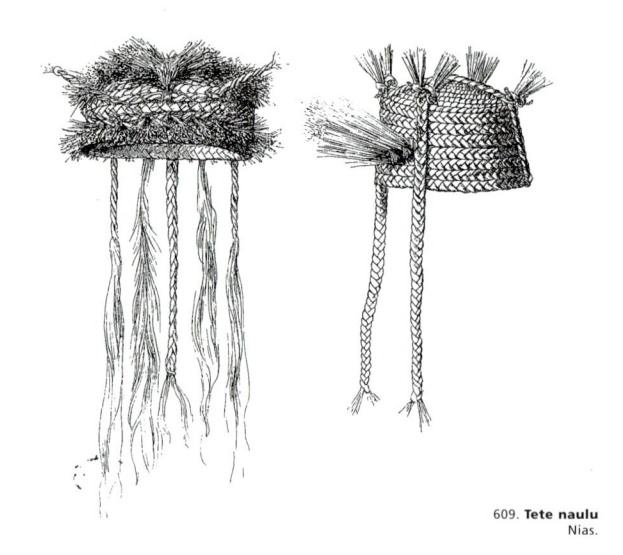

609. **Tete naulu** Nias.

TEUBOEENG TOEMBAK
See *teubueng tumbak*

TEUBUENG TUMBAK
[TEUBOEENG TOEMBAK]
SUMATRA
A spear with an undulating blade.
(ROGERS)

TEUPAT
See *parang teupat*

THIKIN
See *sikin*

THININ
SUMATRA, ACEH
A sword with quite a broad blade and sharpened on one side. The blade's length is c.50 cm.
(DRAEGER)

TIDJARROE
See *tojaru*

TIGAR
JAVA
A sword with a broad, heavy blade. Protruding forged decorations are located on the blade's back and on the edge, near the hilt.
(JASPER AND PIRNGADIE 1930)

TIPU DAYAK
KALIMANTAN
A spear with two points. One point lies in the extension of the shaft. The other, a slightly shorter one, is attached on one side of the longer point. At first it curves out and then up, in a parallel way. Both points have a barb. The story goes that a Dayak attacks with such strength that if a spear enters his body, he is still able to run on and kill the person wielding that spear. The second point of the *tipu dayak* (meaning: 'deceive the Dayak') would allegedly render this impossible.
(GARDNER 1936)

608. **Tete naulu** Nias.

610. **Tigar** Java.

611. **Tipu dayak** Kalimantan.

TIRRIK LADA
SULAWESI
A gadget with which to throw red pepper into the eyes of an adversary, from close range.
(DRAEGER)

TITJIO
See *cabang*

TJABANG
See *cabang*

TJALUK
See *caluk*

TJALUK LAPAR
See *caluk lapar*

TJANDONG
See *rudus*

TJELURIT
See *celurit*

TJETJERET
See *pacret*

TJINANGKE
See *cinanke*

TJIO
See *cio*

TJO JANG
See *co jang*

TJOEDRE
See *cudre*

TJOENDRIK
See *chunderik*

TJOERIGA
See *luju celiko* or *parang rantai*.

TJOK IJANG
See *co jang*

TJOK JANG
See *co jang*

TJOK YANG
See *co jang*

TJOLI
See *coli*

TJOLIKA
See *coli*

TJOMBONG
See *combong*

TJONDRE
See *chunderik*

TJOQ JANG
See *co jang*

TJORIK
See *rudus*

TJOT JANG
See *co jang*

TJUNDRIK
See *kudi*

TO
JAVA, MADURA
A pair of short swords with a parrying bar and a hand protector.
(DRAEGER)

TO SANGTO
JAVA, MADURA
A pair of long, slightly curved swords.
(DRAEGER)

TOA
See *stick swords*

TODI
MADURA
A weapon with numerous shapes and sizes.
(DRAEGER)

TODJARROE
See *tojaru*

TODO
MALUKU, BURU
A sword the blade of which broadens towards the point. The edge and the back are straight, the back curves at the point towards the edge. The scabbard (*katuen* or *todopenan*) is used in the south of Buru as a shield.
(DRAEGER)

TODOPENAN
See *katuen*

TOELEUENG
See *tuleueng*

TOELLANGA
See *tullanga*

TOELON
See *tuleueng*

TOEMBA
See *tumba*

TOEMBA DJANGAT
See *tumba jangat*

TOEMBA MEUDJANGGOT
See *tumba meujanggot*

TOEMBA MEULINGKO
See *tumba meulingko*

TOEMBAK
See *tumbak*

TOEMBAK BOENGONG DJEUMPA
See *tumbak bungong jeumpa*

TOEMBAK LHEE SAGOE
See *tumbak lhee sagu*

TOEMBAK MEULINGKOH
See *tumbak meulingkoh*

TOEMBAK ON BOELOH
See *tumbak on buloh*

TOEMBAK SERAMPANG
See *sarampang*

TOEMBAK TODOEK
See *tumbak toduk*

TOEMBAQ
See *tumbak*

TOEMBOEK LADA
See *tumbok lada*

TOEMBOEQ LADA
See *tumbok lada*

TOEMBUK LADA
See *tumbok lada*

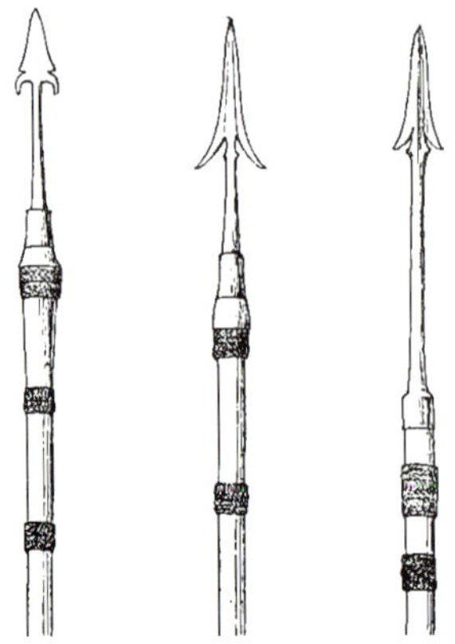

TOHO
NIAS
A collective term for a variety of spears. The point varies from small to large, with near the base two more or less protruding points, or a barb on one side. The shaft's upper side usually has a conical ring of brass or iron and may be decorated with pig's or goat's hair.
(FELDMAN 1990; STONE)

612. **Toho** Nias.

613. **Toho** Blade of illus. 615. L. 25.5 cm.

614. **Toho** Nias.

615. **Toho** Nias. L. 201 cm.

TOHOK
A javelin to which a rope is tied so that it can be retrieved.
(GARDNER 1936)

TOJARU
[TIDJARROE, TODJARROE]
SULAWESI
A spear with a *banranga* (Buginese) or *banrangang* (Makassar): a decoration near the shaft's end, made of horse's or male goat's hair or cock feathers. A cover to protect this decoration is called *sepu-banranga* or *sampu-banranga* (Buginese) or else *purukang-banrangang* (Makassar). The mount is named *pando* (both Buginese and Makassar).
(MATTHES 1874, 1885; SCHRÖDER)

TOLAKI
See *labo*

TOLOGOE
See *tologu*

TOLOGU
[TOLOGOE]
SOUTH NIAS
A sword with a straight-backed blade. The edge is slightly convex. The back runs at the point in a slightly S-shaped curve to the edge. Towards the point the blade broadens. The hilt turns near the top at almost a right angle, and is usually carved in the shape of a monster's head. At the blade the hilt has a brass sleeve ending in a broader rim. The scabbard's top may have a basket filled with amulets. See also *balato*.
(FISCHER)

TOLOR
[TELENGA]
KALIMANTAN, DAYAK
The quiver in which the Dayak carry the poisonous darts for blowpipes. The *tolor* is made of a piece of bamboo. A partition between the segments forms the bottom. The lid is likewise made of a similar piece of which a partition forms the upper part, or is made of wood in the form of a cone or a semi-circular upper side. A wooden girdle hook is attached to the quiver by means of woven strips of rattan. See also *tavang*.
(STONE)

616. **Tojaru** Sulawesi.

617. **Tolor** Kalimantan.

618. **Tolor** Kalimantan. Quiver with *hung* for the pith butts of the arrows.

619. **Tolor** Kalimantan, south east coast. Quiver with iron belt hook.

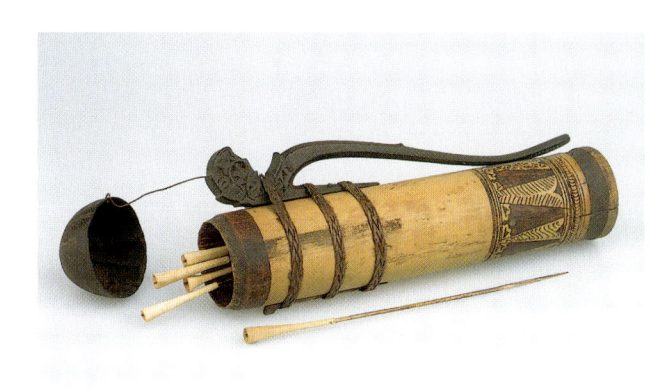

620. **Tolor**
Kalimantan. RMV 2292-93. Acq. in 1935 from G.L. Krol. H. 31.5 cm.

621. **Tolor with hung for the back ends of the blowpipe darts**
Kalimantan. H. 33 cm.

622. **Tolor**
Kalimantan. H. 33 cm.

TOMBAK
See *tumbak*

TOMBAK LADA
See *tumbok lada*

TOMBOK LADA
See *tumbok lada*

TOMBOLADA
See *lopah petawaran*

TONDA
SUMBAWA
An oval shield, the top of which ends in a point. The shield has a grip at the back in the middle.
(DRAEGER)

TONGKAT
TALAUD
A type of short stick or club.
(DRAEGER)

TONKAT PEMUKUL
FLORES
A hardwood stick measuring c.140 cm long and 5 cm in diameter.
(DRAEGER)

TOPO
See stick sword.

TORDJONG
See *lopah petawaran*

TOYA
[TOYAH]
A stick, usually made of rattan, measuring between 150-180 cm and with an overall diameter of 3-5 cm. The *toya* can be held either in the centre or at the end.
(DRAEGER; GARDNER 1936)

TOYAH
See *toya*.

TOYAK
BALI
A weapon strongly resembling a halberd with a plain broad and slightly curved blade.
(DRAEGER)

TRABAI
See *kliau*

TRABAI KLIT KLAU
KALIMANTAN, SEA DAYAK
A *trabai klit klau* is a long shield with a grip, made of a single piece of wood. Its straight sides end in a triangle on the lower and upper sides. It is slightly convexly curved lengthwise with an elevated back in the centre and is carried in the left hand at some distance from the body. This shield is not meant to catch spears, but to change their direction with a flick of the wrist and thus render them harmless. It is often coloured with red ochre or decorated with a carefully executed, intricate pattern. The *trabai klit klau* is large enough for its purpose, but small in comparison with the shields of other peoples. When made of wattled bamboo, it is called *trabai temiang*.
(ROTH 1896A,B)

TRABAI TEMIANG
KALIMANTAN, DAYAK
A variant of the *trabai klit kau*, made of wattled bamboo.
(ROTH 1896A,B)

TRISULA
[BESI TIGA, TERISULA]
JAVA
An arrow of the Javanese deities, demigods and heroes from antiquity. The point has a trident consisting of a crescent with in the centre a straight point. The term *trisula* is also applied to a spear with a trident as point.
(GARDNER 1936; JESSUP; RAFFLES; STONE)

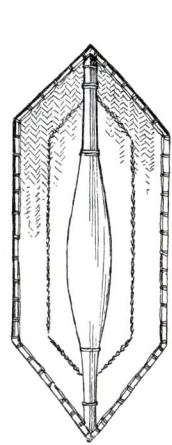

623. **Trabai temiang**
Kalimantan, Dayak.
Bamboo shield.

624. **Trisula**
Java. Spear head with gold encrustation.

625. Trisula
Java. RMV 704-3. Pamor-iron, encrusted with gold and decorated with diamonds. Acq. in 1889. L. 229 cm.

TUDONG
A face mask, part and parcel of combat dress.
(GARDNER 1936)

TUKAK
See *ranjau*

TULEUENG
[TOELEUENG, TOELON, TULON]

SUMATRA

A type of bone, used for making hilts. Known as *tuleueng* (Aceh) and *tulon* (Gayo).
(KREEMER)

TULLANGA
[TOELLANGA]

CENTRAL SULAWESI, TOPANTUNUASU

A spear made of bamboo (*Bambusa longinodes*).
(LEENDERTZ)

TULON
See *tuleueng*

TULUP
A blow-pipe used for firing poisonous darts (*paser*) and made of a straight piece of wood or reed, measuring *c.*150 cm. The *tulup* has no spear-head.
The blow-pipe used for firing bullets has the same name, but does have a spear-head and is made of iron-wood.
On Java the *tulup* has not been in use for centuries. In literature dating from 1598, we read a story about an attack on the Dutch, who were welcomed with a shower of darts. Under the fish tooth tips of these darts two incisions were made, causing them to break after hitting their targets.
On other islands, including Kalimantan, blow-pipes are found to the present day.
(LOEBÈR 1928; RAFFLES 1817A,B; STONE)

TULUPAN
JAVA

A blow-pipe made of bamboo with long segments.
(STONE)

TULWAR
See *podang*

TUMBA
[TOEMBA]

SUMATRA, ACEH, FLORES

A spear, sometimes with a golden or silver shaft (*teubueng*). On Flores these spears are made from a single piece of wood.
(STONE; VELTMAN)

TUMBA JANGAT
[TOEMBA DJANGAT]

A ceremonial spear with a straight blade, enclosed in a brass ornament with vertical protrusions on both sides.
(STONE)

TUMBA MEUJANGGOT
[TOEMBA MEUDJANGGOT]

SUMATRA, ACEH

A lance with a short, broad blade with at the base, on both sides, a small protrusion. The shaft has a circle of large feathers directly under the blade. Only the *sultan* and the three *panglima*s *sagi* were allegedly permitted to possess such lances.
(KREEMER)

TUMBA MEULINGKO
[TOEMBA MEULINGKO]

SUMATRA, ACEH

A thrusting lance with an undulating blade.
(KREEMER)

TUMBAK
[GANDJUR, TOEMBAK, TOEMBAQ, TOMBAK, TUMBAK GANDJUR, WAHOS]

A collective term for spears found all over the archipelago. A main division can be made between spears with straight points (*tumbak bener*) and those with undulating points (*tumbak luk* or *tumbak leres*). Spears have an almost endless variety of figurative forms.
The parts of the *tumbak* are:
(a) a scabbard (*wrongka*), sometimes with a metal strip (*suh*);
(b) the round base of the blade of the *tumbak* (*metuk*);
(c) the shaft (*landejan*);
(d) a ring (*godi*) placed around the shaft. If made of metal, it is called *blongsong* or *srubungan*;
(e) two narrower strips above and below the *godi*;
(f) the shaft's base (*tunjung*);
(g) a ring (*sopal*) around the shaft, just above the tunjung.
(GRONEMAN 1904B; JASPER AND PIRNGARDIE 1930; SNOUCK HURGRONJE 1892; STONE)

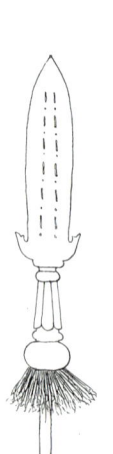

626. Tumba meujanggot
Sumatra, Aceh.

TUMBAK BUNGONG JEUMPA
[TOEMBAK BOENGONG DJEUMPA]

SUMATRA, ACEH

A spear with a short point.
(SNOUCK HURGRONJE 1892)

TUMBAK GANDJUR
See *tumbak*

TUMBAK LHEE SAGU
[TOEMBAK LHEE SAGOE]

SUMATRA
A spear with a triangular point.
(SNOUCK HURGRONJE 1892)

TUMBAK MEULINGKOH
[TOEMBAK MEULINGKOH]

SUMATRA
A spear with an undulating blade.
(ROGERS)

TUMBAK ON BULOH
[TOEMBAK ON BOELOH]

SUMATRA, ACEH, SAMALANGA, PEUSANGAN
A spear, the blade of which is broad in the middle.
(SNOUCK HURGRONJE 1892)

TUMBAK TODUK
[TOEMBAK TODOEK]

A variant of the *tumbak* with barbs on both sides of the point.
(STONE)

TUMBOK LADA
[BOLADO, PEUNOEWA, PISO SUKUL GADING, TEMBOK LADA, TOEBOEK LADA, TOEMBOEQ LADA, TOEMBUK LADA, TOMBAK LADA, TOMBOK LADA]

SUMATRA, MALAYAN PENINSULAR
A dagger with a slightly curved blade. Its edge is situated on the concave side. At the point the edge curves towards the back. The blade resembles the blade of the *sewar*, but is broader and thicker, with along the back usually one or more grooves lengthwise. The length is mostly c.15-20 cm but may reach 45 cm.
The name (*tumbok*, meaning: 'to grind or to crush'; *lada*, meaning: 'pepper') is derived from the hilt's shape which resembles the pounder used to crush pepper-corn. The hilt is short, thick and lies in the blade's extension. The Gayo term is *rudus tarah baji* (meaning: 'cutting of wedges').
In Minangkabau and Negri Sembilan, the *tumbok lada* is the ceremonial weapon of lower chiefs. In Perak and Selangor, the *tumbok lada* is smaller and is considered inferior, a 'woman's weapon'. Because of its size, it can easily be concealed and used unexpectedly.
(FISCHER; GARDNER; KRUIJT; MÜLLER; ROGERS; STONE; VOLZ 1909)

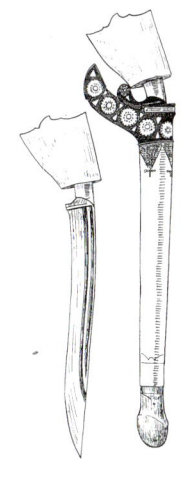

627. Tumbok lada Sumatra.

628. Tumbok lada Sumatra, Karo Batak. L. 30.5 cm.

629. Tumbok lada Sumatra. L. 36.5 cm.

630. Tumbok lada Sumatra. L. 25.5 cm.

631. Tumbok lada Sumatra. L. 25.5 cm.

TUMPULING
JAVA
A spear with one or more barbs on its point.
(STONE)

TUPA
MALUKU
A wooden spear.
(DRAEGER)

TURDJANG
See *turjang*

TURI
See *o humaranga*

TURJANG
[TURDJANG]

SUMATRA, BATAK
A decorative knife with a blade resembling that of a *rencong*, but with another type of hilt.
(VOLZ 1909)

uw

ALPHABETICAL SURVEY

632. War jacket
Nias. RMV 985-1. Acq. in 1894 from Jhr. mr. A.P.C. van Karnebeek. L. 62 cm.

UBLAKAS
See *parang upacara*

ULTUP
[ELTOEP, OELTOEP]

SUMATRA, BATAK

A blow-pipe made of bamboo, mainly used for hunting, with or without a mouth piece. The *ultup* is the national weapon of the Lubus. It has a bamboo exterior. On the inside a second thinner piece of bamboo serves as a lining. The Karo name for this item is *eltep*.
(FISHER 1912; STONE)

ULU
A Malayan name for a hilt.

ULU-ULU
See *tandu tandu*

UMBAN TALI
A sling made of fibres or sometimes of human hair.
(GARDNER 1936)

UMBUL UMBUL
JAVA

A banner made of palm leaf. On one side it is cut off close to the central vein. On the other side it is cut off right against the central vein, leaving behind a streamer at the top.
(STONE)

UPAS
See *ipoh*

UTAH
A round wooden shield, sometimes decorated with metal plates.
(GARDNER 1936)

UTAH-UTAH
A small wooden shield with bells and banners used in war dances.
(GARDNER 1936)

UTAP
[PAPATN, PRONGIN]

KALIMANTAN, DAYAK, BIDAYUH

A long shield varying in shape. Both ends may be rounded, but one rounded and one pointed end also occur. Furthermore, its shape may be octagonal whereby both sides show a slight angle near the centre. Both points end in a roughly rectangular tip. The *utap* may be made of wood (where the handle is carved in one piece along with the shield), tree-bark strengthened with wood or rattan, or wood covered with rattan plait work. The lotus motif can be used as a decoration.

The *utap* was allegedly the original shield of the Sea Dayak, not in use since the beginning of the 20th century. It was made of rattan woven strips strengthened lengthwise with a carved wooden board. The Bidayuh use the term *prongin* for this *utap*.
(CHIN; HOSE; ROTH 1896; SELLATO; STONE)

WADJOE-RANTE
See *baju rantai*

WAHOS
See *tumbak*

WAJU RANTE
See *baju rantai*

633. **Utap**
Kalimantan. RMV 1525-10. Acq. in 1906 from Lansberge Heirs. L. 59 cm.

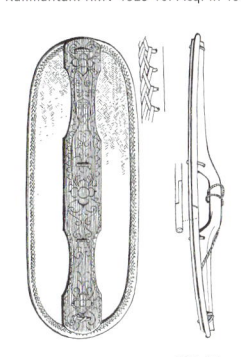

634. **Utap**
Kalimantan, Dayak. L. 51 cm, W. 18 cm.

WALI
SUMATRA, PALEMBANG

A knife with a curved blade used to split rattan.
(MUSEUM 1965)

WANGKIL
JAVA

A machete, mainly used as an agricultural tool. It has a short, broad almost square blade with a long straight hilt about twice as long as the blade.
(RAFFLES; STONE)

635. **Utap**
Kalimantan. RMV 709-1. Acq. in 1889. L. 67 cm, W. 22.5 cm.

also of metal, woven *ijuk* fibre, bark cloth covered with *aren* palm fibre or cloth covered with 'scales' of bark. On Central Sulawesi imitations of European coats of mail are found, but made of small strips of buffalo hide sewn together. Cuirasses of thin iron and shirts covered with scales are found here, too. Apart from these indigenous examples, cuirasses of European origin were also used. These were imitated on Maluku as well. On Flores the foremost fighters wear a type of cuirass made of buffalo hide and decorated with shells. Such combat dresses are also found on the Tanimbar Islands. On Sumatra amongst the Gajo sometimes cushioned combat jackets are worn made from the scaly skin of an armadillo, or from thick tree bark covered with fish scales, such as of the scaroid (*Pseudosarius marine*). These are attached to the underlying material with thin rattan strips, or cord. Amongst the Murut cushioned combat jackets are used, covered with *cowrie* shells.

(DRAEGER; FELDMAN; FURNESS; GRUBAUER; KEPPEL 1846A, 1846B; ROTH 1896B; STONE; VOLZ 1912; VOSKUIL)

639. War jacket
Sulawesi, Toraja.

WARAJANG
JAVA
A mythical arrow used by gods. It has a thin, undulating tip or a broad triangular point with barbs.
(RAFFLES; STONE)

WEDOENG
See *wedung*

WEDONG
See *wedung*

WEDUNG
[WEDOENG, WEDONG]
JAVA, BALI
A short, broad machete. Its blade has a straight back and an S-shaped edge. It may be made of smooth iron, but *pamor* forge work also occurs. The back is sometimes sharpened along c.1/3 of its length from the point. The blade's base is straight and stands at an angle of 90° to the back. This base usually has decorations shaped as filed-out indentations or small teeth (*greneng*). A round 'eye' or hole (*kembang kacang*) is sometimes found. Furthermore, the base may be decorated with inlay work and representations of the mythical snake (*naga*), leaf and floral motifs. The tang (*peksi*) is made out of the blade's thick back rim. Between the spike and the blade, a pentagonal segment (*metok*) is forged. The short hilt,

636. War jacket
Seram. RMV 300-1784. Acq. in 1878 World Fair, Paris. L. 124 cm, W. 34 cm.

WAR JACKET
War jackets are found all over the Indonesian archipelago in every shape and size. Especially on Nias a coat of mail or cuirass is a fixed part of the combat dress and is known there under various terms. Usually they are made of hard leather, but

637. War jacket
Nias. RMV 985-1. Acq. in 1894 from Jhr. mr. A.P.C. van Karnebeek. L. 62 cm.

638. War jacket
Nias. Warriors wearing a war jacket, a *baluse* (shield) and a *toho* (spear).

640. War jacket
Timor-Laout. Warrior wearing a war jacket.

which is pentagonal on Java, forms as it were, an extension of the *metok*. The hilt is usually made of wood, but other materials such as animal tooth are also found. Its smooth upper part is flattened. The *wedung* on Bali has a narrower shape and is usually more lavishly decorated. On the rear near the base a lengthy decoration is sometimes added. The hilt and the *metok* of the Balinese *wedung* are instead of pentagonal, usually round in cross-section. It is sometimes chiselled. The wooden scabbard usually follows the blade's shape, but may have a rectangular lower part. Its mouth has a broadened rim all around. At the rear side the scabbard has a large horn hook shaped resembling a shoe-horn, longer than the scabbard, with which the *wedung* can be hitched onto the belt. This hook is attached to the scabbard by means of strips of horn, rattan or metal. On expensive examples the scabbard sometimes has one or two round or comma-shaped golden or silver mounts. The *wedung* is carried in the palace (*kraton*) as a symbol of servitude to the *sultan* for performing such tasks as the cutting of shrubs (*belukar*) or even the most humble work such as grass cutting. It is, therefore, no longer a real weapon but rather a work-tool carried on the left hip and used as personal decoration.

(EGERTON; GARDNER 1936; GRONEMAN 1910A, 1904B; JASPER AND PIRNGADIE 1930; PARAVICINI; RAFFLES 1817A; SCHMELTZ 1890; STONE; VOSKUIL 1921)

641. Wedung
North Java. RMV 963-4. Acq. in 1893 from Baron mr. L.A.J.W. Sloet Van Der Beele Van Nispen. L. 34.5 cm.

642. Wedung
Java. Rear view of illus. 641.

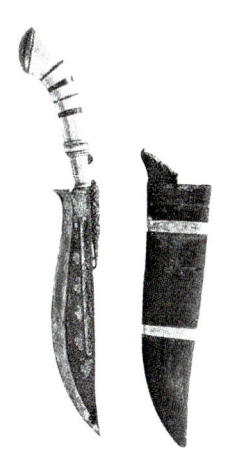

643. Wedung
Bali. L. 32 cm (in scabbard).

644. Wedung
Bali, Singaraja. Gold encrustation.

645. Wedung
Java. A Javanese in the court dress, wearing a *wedung* and a *keris*.

646. Wedung
Java. L. 31 cm.

647. Wedung
Bali. L. 32.5 cm.

WUHU
FLORES
A bow of various sizes.
(DRAEGER)

INDEX OF NAMES OF WEAPONS AS FOUND ON ISLAND/GROUP OF ISLANDS

ADONARA
STICK SWORD

ALOR
MOSO
RUGI

AMBON (MALUKU)
SALAWAKU

BALI
ARIT
ARIT MENGOBED
BESI DUA
BESI LIMA
CABANG
CEREMONIAL AXE
KERIS
LANGKAP
PARANG BENGKOK
PEDANG BENTOK
PEDANG CHEMBUL
PRINGAPUS
PUNGLU
TAJI
TEKKEN
TOYAK

BANDA ARCHIPELAGO
KISA
RAA

BURU (MALUKU)
ENHERO
MAEN
SALAWAKU
SUMPING

ENGGANO
BLUDGEON
EKAJO
EKEH
FEJI
FEJI KAIKBARU
KANAKINIE

FLORES
AGANG
BELIDA
BELO
KORUNG
LABE
NUME
PRICAI KAYU
REPAU
STICK SWORD
SUMARA
TONGKAT PEMUKUL
TUMBA
WUHU

HALMAHERA (MALUKU)
HUMARANGA
MA DADATOKO
O GOHOMANGA MA URU
O HUMARANGA
SALAWAKU

HARUKU (MALUKU)
TABAK

JAVA
ALAMANG
ARIT
BADEK
BANDANG

BANDOL
BANDRING
BARADI
BEDER
BENDO
BESI DUA
BESI LIMA
BODIK
BUGIS
CABANG
CALUK
CAYUL
CIO
COLI
CHACRA
CHAKERA
CHEMETI
DENDA
DIWAL
DWISULA
ENDONG PANA
GADA
GANJING
GENDAWA
GRANGGANG
HALBERD
HONGKIAM KEK
HWA KEK
INDAN
JAMBIAH
KERIS
KERIS MAJAPAHIT
KIAM BOKIAM
KOWLIUM
KUDI WAYANG
KUJUNGI
KWAI
LAJATANG
LARBANGO
LARBIDO
LAWI AYAM
LIANGTJAT
LUWUK
LUKI
MENTAWA
MENTOK
PANA
PARANG BENGKOK
PARANG PANJANG
PASER
PASPATI
PATREM
PEDANG BENTOK
PEDANG LURUS
PEDANG SUDUK
PISO BELATI
RODA DEDALI
RONKEPET
RONTEGARI
ROTI KALONG
RUDING LENGON
SEGU
SEPA
TAMENG
TAWOK
TEKKEN
TELABUNA
TELEMPANG
TIGAR
TO
TO SANGTO
TRISULA
TULUPAN
TUMPULING
UMBUL-UMBUL
WANGKIL

WARAJANG
WEDUNG

KALIMANTAN
BADIQ LOKTIGA
BAJU EMPURAU
BAKIN
BARONG
BAYU
BELANGAH
BERANDAL
BLASUNG
BUKO
CHANDONG
COAT OF MAIL
DOHONG
DUKN
GAGONG
GRANGGANG
HAUT NYU
HUNG
ISAU
JIMPUL
JOWING
KALAVIT
KALIHAN
KALUPU
KAMPILAN
KARIS
KARUNKUNG
KATAPU
KATUPU KALOI
KERIS
KLAMBI TAYAH
KLAUBUK
KLEBIT BOK
KLIAU
KOHONG KALUNAN
LADING BELAJUNG LAMAH
LADING CARA
LADING JAWA
LAJAU
LANGA
LANGGAI TINGGAN
LAVONG
LUNJU
LUTONG
MANDAU
MANDAU PASIR
NIABOR
PAKAYUN
PALUON
PANDAT
PARABAS
PARANG BEDAK
PARANG KAJULIE
PARANG LATOK
PARANG NABUR
PARANG NEGARA
PARANG PARAMPUAN
PARANG PATAH
PARANG PEDANG
PATOBANG
PEDANG BERANDAL
PEDANG JAWIE BESAR
PISO TONGKENG
PUOT
RANDU
RENTENG
SABIT MATA DORA
SADAP
SADOP
SADUP
SAGU SAGU
SAMBILATIUNG
SAMPULAU ANGGANG

SANGKOH
SAPITABON
SARAMPANG
SILIGIS
SINA PAYED
SLIGHI
SOPOK
SUMPITAN
SUNDANG
TALAWANG
TAMBULOH
TANGIRRI
TAVANG
TELEP
TEMBILAH
TEPUS
TIPU DAYAK
TUMBA JANGAT
TOLOR
TRABAI KLIT KLAU
TRABAI TEMIANG
UTAP

KISAR
OPI

LINGGA ARCHIPELAGO
BLUDGEON
SALIGI

LOMBOK
KERIS
PECUT
TAMENG

MADURA
ARIT
ARIT BIASA
ARIT LANCHAR
BADEK
BULU AYAM
CABANG
CALUK LAPAR
CELURIT
CIO
HONGKIAM KEK
HWA KEK
JAMBIAH
KERIS
KIAM BOKIAM
KOWLIUM
KUJUNGI
KWAI
LAJATANG
LARKAN
LIANGCAT
PISO BELATI
TEKKEN
TO
TO SANGTO
TODI

MALUKU (VARIOUS ISLANDS)
BLUDGEON
CABANG
GALA
KATUEN
PEDA
SALAWAKU
TODO
TUPA

MENTAWAI ARCHIPELAGO
BELADAU
BUKBUK
KURABIT

PALITAI
SALUKAT
SIKIM GAJAH
SOAT

NIAS
BALAHOGO LEMA'A
BALATO
BALUSE
BARU LEMA'A
BARU OROBA
BARU SINALI
BULU SEWA
BUM'BERE TOFAO
DANGE
GARI
G-OROBANA'A
NIO
SEIMBU NODA
SEWAH KECIL
SI EULI
SIKIM GAJAH
TAKULA SINALI
TAKULA TOFAO
TETE NAULU
TOHO
TOLOGU

RIAU ARCHIPELAGO
BELADAU
BLUDGEON
GALAGANJAR
PEDANG JENAWI
SALIGI
SIKIM GAJAH

ROTI
TAFA

SAVU
HEMOLA

SERAM (MALUKU)
LOPU
SALAWAKU
SANOKAT

SOLOR
STICK SWORD

SULAWESI
ALAMANG
AWOLA
BADEK
BAJU PA'BARANI
BAJU RANTAI
BALADAU
BALULANG
BANGKUNG
BARONG
BASI AJE TADO
BASI BARANGA
BASI PAKA
BASI PARUNG
BASI SANGKUNG
BASI SANRESANG
BASI TAKANG
BLUDGEON
BOLONGKIDANG
CABANG
CUIRASS
DOKE
DOKE KADANGAN
DOKE LEPANG
DOKE PANGKA
DUA LALAN

INDEX OF NAMES OF WEAPONS AS FOUND ON ISLAND/GROUP OF ISLANDS

GAJANG
HUI THO
INDUPO
KALIJAWO
KALIJAWO MALAMPE
KALIJAWO MALEBU
KAMPILAN
KANJAI
KANTA
KAWALI
KERIS
KLIAU
KORAMBI
LABO
LABO BALANGE
LABO BALE BALE
LABO TO DOLO
LABO TOPANG
LAMENA
LASAG
LIYO LIYO
LI PUN
MATANA KNIFE
MOSO
PADIMPAH
PAMULU
PANGHO
PARANG RANTAI
PARANG UPACARA
PASEKI
PEDA
PENAI
PISO LAMPAKAN
PISU
POKE
RUGI
SALIGI
SAPU
SAPURU
SONGKOK
STICK SWORD
SUMARA
SUMPI
SUNDANG
TAMBENG
TAMBUK
TA MING
TAMPELAN
TANDU-TANDU
TAO
TAWARA
TIRRIK LADA
TOJARU
TULLANGA

SUMATRA
ALI-ALI
AMANREMU
AMBALANG
ANDAR ANDAR
ARIT
BADEH
BADEK
BAJU GUS
BALASAN
BANGKUNG
BAWAR
BELADAU
BEREGU
CHEMETI
CHENANGKAS
CINANGKE
CO JANG
COMBONG
DAMAOQ
DOUBLE KNIFE
DOUBLE SWORD
GADA LIMBANG
GADOOBANG
GALAIJANG TOKONG
GEUDUBANG
GEUREUPOH
GOLOK REMBAU

GONTAR
GUPUK
GUPUK PERAGIT
HAMPANG HAMPANG
HANGAN
HINA
HOJIUR
HUJUR
HULU
JAMBIAH
JONAP DAIRI
JONO
KALASAN
KALASAN SITUKAS
KAMPAR SABIT
KAPAK
KASO
KATUNGUNG
KERIS
KERIS PANJANG
KETUPUNG
KLEWANG CARA ACEH
KLEWANG PUCOK MEUKAWET
KLEWANG TEBAL HUJONG
KORAMBI
KUJUR
KUTA JA
LADIENG
LADINGIN
LAJUK LAJUK
LAMBING
LAWI AYAM
LEMBING RAJA
LEMBING SI DUA DUA
LEMING KAPAK
LEUMBENG
LOMBU LOMBU
LOPAH
LOPAH PETAWARAN
LUJU ALANG
LUJU ALAS
LUJU CELIKO
LUJU MUGENTA
LURIS PEDANG
MERMU
MUNDO
PACRET
PADANG SABIT
PALANGGE
PAMANDAP
PANGUR
PANGUR CUT CUT
PANJI
PARANG CANDONG
PARANG GEDAH
PARANG GEUDANG
PARANG IKU LINONG
PARANG LEUNTEK
PARANG PANJANG
PARANG TEUPAT
PEDANG
PEDANG ABEUSAH
PEDANG BENTOK
PEDANG LHEE KURO
PEDANG MEUTAMPU
PEDANG SIEM
PEDANG TEUNUANG
PEDANG TEUPEH
PELAJU
PEURAWOT
PEURISE
PEURISE AWE
PEURISE KAJEE
PEURISE NILO
PEURISE PARU
PEURISE TEUMAGA
PISO
PISO BELATI
PISO ENGKAT
PISO GADING
PISO HALASAN
PISO MARIHUR
PISO NI DATU

PISO PERLAJO
PISO SANALENGGAM
PODANG
PUE SURING
RAJA DUMPAK
RAUT
RAWET
RAWIT
RAWIT PENGUKIR
RENCONG
RENCONG PUCOQ PAKU
RUDUS
SABIT
SADEUEB
SAKIN
SAMPAK GLIWANG
SEKIN
SEWAR
SIKIM GAJAH
SIKIN IKU MANOQ
SIKIN LAPAN SAGU
SIKIN MEUKSARUEK ULAT
SIKIN PANJANG
SIKIN PASANGAN
SIKIN PREUNGGI
SIMUNUNG
SI OR
SIRAUI
SUKUL
SURIK
TAKA BLADE
TAPAK KUDAK
TARAH BAJU
TEMBULONG
TEUBUENG TUMBAK
THININ
TULEUENG
TUMBA
TUMBA MEUJANGGOT
TUMBA MEULINGKO
TUMBAK BUNGONG JEUMPA
TUMBAK LHEE SAGU
TUMBAK MEULINGKOH
TUMBAK ON BULOH
TUMBOK LADA
TURJANG
ULTUP
WALI

SUMBA
KABEALA
TAMUA

SUMBAWA
KERIS
TAMUA
TONDA

TALAUD ARCHIPELAGO
KAMPILAN
TONGKAT

TANIMBAR
SURUK

TIMOR
BLUDGEON
GADA
HEMOLA
KAHUK
KAHUK ISIN
SUNI
SURIK
TAMENG

WETAR
OPI
RUGI

WEAPONS FOUND IN LARGER PARTS OF THE ARCHIPELAGO AND/OR THE EXACT PROVENANCE OF WHICH IS NOT CERTAIN

BAJU LAMINA
BALING-BALING
BATU RAJUT
CHANDAK
CHELANA
CHEMETI
CHOKUMER
CHUNDERIK
CHURA SI-MANJAKINI
CUDRE
GADA
GAYONG
GEDUBAN KLEWANG
GEGANIT
GOBANG
GOLOK
GOLOK BANGKONG
HELMET
ILANUN KAMPILAN
KALUS
KAPAK JEPUN
KATOK
KECHIL
KECHUBONG
KELAMBU RASUL ALLAH
KLEWANG
KNIFE
KOTAU
KUDI
KUDI TRANCHANG
LADING TERUS
LIDAH AYAM LIPAT
PANAH AYER
PARANG
PARANG BANKONG
PARANG CHAKOK
PARANG CHANDONG
PARANG GABUS
PARANG GINAH
PARANG GONDOK
PARANG KOTENG
PARANG LADING
PARANG LOTOK
PARANG ONGKOK
PARANG PAJAH
PARANG PANCONG
PARANG PANDAH
PARANG PANGGONG
PARANG PENDAK
PARANG PERANGGI
PARANG SA-KAMPOK
PARANG SARI
PEDANG
PEDANG BERTUPAI
PEDANG CHAKOK
PEDANG KASOQ
PEDANG ON JOQ
PEDANG PEMANCHONG
PEDANG PERBAYANGAN
PEDANG RAJA UJONG
PELANTEK
PENUMBAK TEMBAGA
PINANG LAYAR
PISO RAOUT
RANGGAS
RANGIN
RANJAU
SABOK
SANG KAUW
SANGGA MARA
SANGSANG
SELIGI
SENANGKAS BEDOK
SERUNJONG
SHIELD
SIBAK
SIKAPAN

SODAK
SPEAR
SUDUK
SULU KLEWANG
SWORD
TALEMPAK
TALI PEDANG
TARBIL
TEKPI
TEMBONG
TEMBONG KEMBOJA
TOHOK
TOYA
TUDONG
TULUP
TUMBAK
TUMBAK TODUK
UMBAN TALI
UTAH
UTAH-UTAH
WAR JACKET

SELECT BIBLIOGRAPHY
List of Abbreviations

AA	Ars Asiatica
AK	Aziatische Kunst
AKKNH	Annalen des K.K. Naturhistorischen Hofmuseums
BKI	Bijdrage Koninklijk Instituut
BSGAE	Bulletin der Schweizerischen Gesellschaft für Anthropologie und Ethnologie
BTLVNI	Bijdragen tot de Taal-, Land- en Volkenkunde van Nederlandsch Indië
CAT. REM	Catalogus van 's Rijks Ethnographisch Museum
CI	Cultureel Indië
ENI	Encyclopaedie van Nederlandsch-Indië
IAE	Internationales Archiv für Ethnographie
IG	Indische Gids
IMT	Indisch Militair Tijdschrift
JAI	Journal of the Anthropological Institute
JFMSM	Journal of the Federated Malay States Museums
JMBBRAS	Journal of the Malaysian Branch of the British Royal Asiatic Society
JMVL	Jahrbuch des Museums für Volkerkunde zu Leipzig
JRAIGBI	Journal of the Royal Anthropological Institute of Great Britain and Ireland
JSBRAS	Journal of the Straits Branch of the Royal Asiatic Society
KBGKW	Koninklijk Bataviaasch Genootschap van Kunsten en Wetenschappen
KMVV	Königliche Museen für Völkerkunde, Veröffentlichung
MASB	Memoirs of the Asiatic Society of Bengal
MKNAW	Mededeelingen der Koninklijke Nederlandsche Akademie van Wetenschappen
MNMS	Memoirs of the National Museum, Singapore
OZ	Ostasiatische Zeitschrift
PKEMD	Publicationen aus dem Königlichen Ethnographischen Museum zu Dresden
SMJ	Sarawak Museum Journal
TBB	Tijdschrift voor het Binnenlandsch Bestuur
TITLV	Tijdschrift voor Indische Taal-, Land- en Volkenkunde
TKNAG	Tijdschrift van het Koninklijk Nederlandsch Aardrijkskundig Genootschap
TNAG	Tijdschrift van het Aardrijkskundig Genootschap
TNION	Tijdschrift Nederlandsch Indië Oud en Nieuw
VBGKW	Verhandelingen van het Bataviaasch Genootschap van Kunsten en Wetenschappen
VEMA	Vrienden van het Ethnografisch Museum Antwerpen
VKITLV	Verhandelingen van het Koninklijk Instituut voor Taal-, Land- en Volkenkunde
VNN	Verre Naasten Naderbij
ZHWK	Zeitschrift für Historische Waffen- und Kostümkunde

A

ANNANDALE, N.	1905-1907	'Some Malayan Weapons', MASB, I.
ANON.	1938	Atlas van Tropisch Nederland. Batavia.
AVÉ, J.B. AND KING, V.T.	1986	Borneo. Oerwoud in ondergang, Culturen op drift. Leiden.

B

BAKAR BIN PAWANCHEE, A.	1947	'An Unusual Keris Majapahit', JMBBRAS (December Vol.).
BANKS, E.	1940	'The Keris Sulok or Sundang', JMBBRAS (August Vol.).
BARBIER, J.P.	1984	Indonesian Primitive Art from the Collection of the Barbier-Müller Museum, Geneva. Dallas.
BEER, J. DE, DORHOUT, N. AND SCHOEMAKER, M.	1993	Borneo. Dayak en Punan, de inheemse bevolking van Borneo. Delft.
BEIDATSCH, A.	1974	Waffen des Orients. München.
BEZEMER, T.J.	1931	Indonesische Kunstnijverheid. 's-Gravenhage.
BICKMORE, A.S.	1868	Travels in the East Indian Archipelago. London.
BISHOP, C.W.	1938	'Long-Houses and Dragon-Boats', Antiquity, Vol. XII.
BISSELING, G.A.L.	c.1950?	Tekeningen van Indonesische Wapens. Unpublished ms.
BISSELING, G.A.L. AND VERMEIREN, P.	1982	Rentjongs. Antwerpen.
BOCK, C.	1881	The Head Hunters of Borneo. London.
BODROGI, T.	1973	Art of Indonesia. London.
BOSCH, F.D.K.	1948	De Gouden Kiem, Inleiding in de Indische Symboliek. Amsterdam.
BRAKEL, J.H. VAN, et al.	1987	Budaya Indonesia, Kunst en Cultuur in Indonesië. Amsterdam.
BROEK, W.G.	1940	'Bataksche Zwaardgrepen', CI, Vol. 2, Oct./Nov.
BUSCHAN, G., et al.	1923	Australien und Ozeanien Asien. Stuttgart.

C

CABATON, A.	1911	Java, Sumatra and other Islands of the Dutch East Indies. New York.
CATO, R.	1996	Moro Swords. Singapore.
CAYLEY-WEBSTER, H.	1898	Through New Guinea and the Cannibal Countries. London.
CHADWICK, N.J. AND COURTNEY, P.P.	1983	Punan Art and Artefacts. Townsville.
CHIN, L. AND MASHMAN, V.	1991	Sarawak, Cultural Legacy, a living Tradition. Kuching.
CLERCQ, F.S.A. DE	1890	Bijdragen tot de kennis der residentie Ternate. Leiden.
COLLET, O.J.A.	1925	Terres et Peuples de Sumatra. Amsterdam.
COPPENS, F., et al.	1999	Dodenritueten en Koppensnellerij. De Cultuur van de Dayak op Borneo. Sint-Niklaas.
COVARRUBIAS, M.	1965	The Island of Bali. New York.
CRIBB, R.	2000	Historical Atlas of Indonesia. Honolulu.

D

DIELES, J.G.	1980	Blanke Wapens uit de Gordel van Smaragd. Amsterdam.
DONGEN, P.L.F. VAN, et al.	1987	Topstukken uit het Rijksmuseum voor Volkenkunde / Masterpieces from the National Museum of Ethnology. Leiden.
DRAEGER, D.F.	1972	Weapons and Fighting Arts of the Indonesian Archipelago. Tokio.
DUUREN, D.A.P. VAN	1996a	De Kris. Amsterdam.
	1996b	'Krissen in het Rijksmuseum te Amsterdam', AK, Vol. 26, Nr. 2.
	1998	Krissen. Een beredeneerde bibliografie. Amsterdam.

E

EGERTON OF TATTON, LORD	1896	A Description of Indian and Oriental Armour. London [Repr.: Harrisburg, 1968].
ENGEL, J.	1980	Geschiedenis en Algemeen Overzicht van de Indonesische Wapensmeedkunst. Amsterdam.
EVANS, I.H.N.	1929	'Type of Keris Sampir and Hilts in the Malay States, Java and Celebes', JFMSM, XII.
	1932	'A Note on the Kingfisher Keris, JFMSM, XV.

F

FABER, P., et al.	1987	Schatten van het Museum voor Volkenkunde Rotterdam. Rotterdam.
FELDMAN, J.A.	1985	The Eloquent Dead. Ancestral Sculpture of Indonesia and Southeast Asia. Los Angeles.
FELDMAN, J.A., et al.	1990	Nias Tribal Treasures. Delft.
FISCHER, H.W.	1903	'Iets over de wapens uit de Mentawei-verzameling van 's Rijks Ethnographisch Museum te Leiden', IAE, XVIII.
	1909	Cat. REM, IV, De Eilanden om Sumatra. Leiden.

	1912	Cat. REM, VI, Atjèh, Gajo- en Alaslanden, Sumatra I. Leiden.		1908	'Nog iets over Messinghelmen, -schilden en de Pantsers in het Oostelijke deel van de O.I.-Archipel', IAE, XVIII.
	1914	Cat. REM, VIII, Bataklanden, Sumatra II. Leiden.	HOF, G.R.	1967	'De Krissen van Indonesië'. De Wapenverzamelaar, Jaargang 5.
	1916	Cat. REM, X, Midden-Sumatra, Sumatra III. Leiden.	HOLSTEIN, P.	1930	Contribution à l'Etude des Armes Orientales, Inde et Archipel Malais. 2 Vols. Paris.
	1918	Cat. REM, XII, Zuid-Sumatra, Sumatra IV. Leiden.	HOOP, A.N.J.TH. VAN DER	1939	'De ethnographische verzameling', KBGKW, Jaarboek VI.
	1920	Cat. REM, XIV, Sumatra-Supplement. Leiden.		1949	Indonesische Siermotieven, Bandoeng.
FISCHER, H.W. AND RASSERS, W.H.	1924	Cat. REM, XVII, De oostelijke Kleine Soenda-eilanden. Leiden.	HOSE, C. AND MCDOUGALL, W.	1912	The Pagan Tribes of Borneo. Vol. I. London.
FORMAN, W.B. AND SOLC, V.	1958	Schwerter und Dolche Indonesiens. Prag.	HOSE, C.	1926	Natural Man, A Record from Borneo. London. [Repr.: Singapore, 1988].
FOY, W.	1899	'Schwerter von der Celébes See', PKEMD, Band XII.	HUBBACK, T.R.	1905	Elephant and Seladang Hunting in Malay. London.
FREY, E.	1986	The Kris, Mystic Weapon of the Malay World. Singapore.	HUYSER, J.G.	1916-1917	'Het Vervaardigen van Krissen', TNION.
FURNESS, W.H.	1902	The Home-Life of Borneo Head-Hunters. Philadelphia.		n.d.	'Een Oud-Javaansch Djimat-Wapen', TNION.
GARDNER, G.B.	1933	'Notes on Two Uncommon Varieties of the Malay Keris', JMBBRAS, XI (Dec. Vol.).	INDONESISCH ETHNOGRAFISCH MUSEUM	1974	Sumatraanse Schoonheid. Delft.
	1936	Keris and Other Malay Weapons. Singapore.[Repr.: Wakefield, 1973].			
GIMLETTE, J.D.	1923	Malay Poisons and Charm Cures.	JACOBS, J.	1894	Het Familie- en Kampongleven op Groot-Atjeh. Deel 2. Leiden.
GOETZ, H.	1927	'Beiträge zur Indischen Waffenkunde I, zur Geschichte des Javanischen Kris', ZHWK, Neue Folge, II.	JACOBSEN, A.	1896	Reise in die Inselwelt des Banda-Meeres. Berlin
GOMES, E.H.	1911	Seventeen Years among the Sea Dyaks of Borneo. London.	JANSE, H.J.	1971	'Enkele Kanttekeningen bij een Javaanse Kris', Armamentaria, Nr. 6.
GREVE, R.G.	1992	Keris. Beschouwingen en verhalen over de krissen van Indonesië. Driebergen - Rijsenburg.	JASPER, E.F.	1904	'Inlandsche Methoden van Hoorn-, Been-, Schildpad-, Schelp- en Paarlemoerbewerking', TITLV, Deel XLVII.
GRIFFITH-WILLIAMS, G.C.	1937	Suggested Origin of the Malay Kris. Singapore.		1906	'Het Eiland Bawean en zijn Bewoners', TBB, Nr. 31.
GROENEVELDT, W.P.	1960	Historical Notes on Indonesia and Malaya compiled from Chinese Sources. Djakarta.	JASPER, J.E. AND PIRNGADIE, M.	1930	De Inlandsche Kunstnijverheid in Nederlandsch Indië. Deel 5. Den Haag.
GRONEMAN, I.	n.d.	'Het Njirami of de Jaarlijksche Reiniging van de Erfwapens en andere Poesaka's', IAE, Band 17.	JENSEN, K.S.	1998	Den Indonesiske Kris, et symbolladet vaben. Vaabenhistoriske Aarborger, Nr. 43. Copenhagen.
	1904a	'Pamor en Pamormotieven', Weekblad voor Indië.	JESSUP, H.I.	1990	Court Arts of Indonesia. New York.
	1904b	'Nikkelpamor', Weekblad voor Indië, nr. 24.	JOCHIM, E.F.	1893	'Beschrijving van den Sapoedi Archipel', TITLV, Nr. 36.
	1904c	'Over Pamor en Pamorsmeedkunst', Weekblad voor Indië.	JONGE, N. DE (ED.)	1990	Indonesia, apa kabar? Meppel.
	1904d	'Pamor-Loewoe en nog wat', Weekblad voor Indië, nr. 42.	JONGE, N. DE AND DIJK, T. VAN	1995	Tanimbar. De unieke Molukken-foto's van P. Drabbe. Amsterdam.
	1904e	'Pamor-Smeedkunst', De Java Bode, 29.6.1904.	JUYNBOLL, H.H.	1909	Cat. REM, Deel I, Borneo, I. Leiden.
	1904f	'Nikkel als Pamor', De Java Bode, 12.7.1904.		1909	Cat. REM, Deel V, Javaansche Oudheden. Leiden.
	1906	'De Vorderingen der Pamor-Smeedkunst', Jogjakarta, 13.8.1906.		1910	Cat. REM, Deel II, Borneo II. Leiden.
	1908	''t Behoud en de Herleving van een echt Javaansche Kunst', Dagblad De Locomotief, 3.12.1908.		1912	Cat. REM, Deel VII, Bali en Lombok. Leiden.
				1914	Cat. REM, Deel IX, Java I. Leiden.
	1910a	'Der Kris der Javaner', IAE, Band 19, 21.		1916	Cat. REM, Deel XI, Java II. Leiden.
	1910b	Pamor-Wapens. Jogjakarta.		1918	Cat. REM, Deel XIII, Java III. Leiden.
GRUBAUER, A.	1913	Unter Kopfjägern in Central-Celebes. Leipzig.		1920	Cat. REM, Deel XV, Java IV. Leiden.
				1922	Cat. REM, Deel XVI, Celebes I. Leiden.
				1925	Cat. REM, Deel XVIII, Celebes II. Leiden.
HAMEL, J.	1968-1971	'Javaanse Krissen', Budidaja, Vol. 1, Nrs. 3, 4; Vol. 2, Nr. 1; Vol. 3, Nr. 1.		1927	Cat. REM, Deel XIX, Celebes III. Leiden.
HAMID, P., et al.	1990	Senjata tradisional Daerah Sulawesi Selatan. Proyek Inventarisasi dan Pembinaan Nilai-Nilai Budaya Departemen Pedidikan dan Kebudayaan Direktorat Jenderal Kebudayaan Direktorat Sejarah dan Nilai Trasdisional.		1928	Cat. REM, Deel XX, Philippijnen. Leiden.
				1930	Cat. REM, Deel XXI, Molukken I. Leiden.
				1931	Cat. REM, Deel XXII, Molukken II. Leiden.
				1932	Cat. REM, Deel XXIII, Molukken III. Leiden.
HARSRINUKSMO, B. AND LUMINTO, S.	1988	Ensiklopedi Budaya Nasional, Keris dan senjata tradisional Indonesia lainnya. Jakarta.	KALFF, S.	1923	'Javaansche Poesaka', Djawa, III.
			KARIM, N.	1993	Senjata tradisional masyarakat Daerah Jambi. Proyek Penelitian Pengkajian dan Pembinaan Nilai-Nilai Budaya.
HAZEU, G.A.J.	1904	'Iets over Koedi en Tjoendrik', TITLV, Nr. XLVII.			
HEEKEREN, H.R. VAN	1958	The Bronze-Iron Age of Indonesia. VKITLV, Deel XXII.	KAT ANGELINO, P. DE	1921	'Over Smeden en eenige andere Ambachts-lieden op Bali', TITLV, Deel 60, Afl. 3 en 4.
HEIN, A.R.	1890	Die Bildende Künste bei den Dayaks auf Borneo. Wien.	KEITH, H.G.	1938	'Keris Measurements from North Borneo', JMBBRAS (July Vol.).
HEIN, W.	1899	'Indonesische Schwertgriffe', AKKNH, Band XIV.	KEPPEL, H.	1846a	The Expedition to Borneo of H.M.S. Dido. Vol. I, London.
HEINE-GELDERN, R.	1932	'Über Kris-Griffe und ihre mythischen Grundlagen', OZ, Neue Folge, Jahrgang 8.		1846b	The Expedition to Borneo of H.M.S. Dido. Vol. II, London.
HELFRICH, O.L.	1888	'De Eilandengroep Engano', TKNAG, Tweede Serie, Deel V.	KERNER, M.	1996	Keris-Griffe aus dem malayischen Archipel. Zürich.
HENDRIKS, A.	1842	'Iets over de wapenfabricatie op Borneo', VBGKW, 18.	KOESOEMO, S.D.	1906	'Dari hal sendjata orang Madoera', TBB, 31.
HEYST, A.F.C.A.VAN	1942	'Anggang Gading', CI, Vol. 4, Jan./Feb.	KOL, H.H. VAN	1914	Driemaal dwars door Sumatra en zwerftochten door Borneo. Rotterdam.
HILKHUIJSEN, J., LEATOMU, D. AND WASSINK-VISSER, R.	1981	Pameran masohi maluku; De Molukken tussen traditie en toekomst. Delft.	KOPPESCHAAR, C.	1977	'Een Kris laat niet met zich spotten', KIJK, Augustus.
HILL, A.H.	1956	'The Keris and other Malay Weapons', JMBBRAS, Vol., 29, Part 4, No. 176.			
	1962	The Malay Keris and other Weapons. Singapore.			
HOËVELL, G.W.W.C. VAN	1908	'Der Kris von Süd-Celebes', IAE, XVIII.			

KOTTEN, B.K.	1990-1991	Senjata Tradisional Daerah Nusa Tenggara Timur. Proyek Inventarisasi dan Pembinaan Nilai Nilai Budaya Daerah Direktorat Sejarah dan Nilai Tradisional Departemen Pedidikan dan Kebudayaan.
KREEMER, J.	1922	Atjèh. I, Leiden.
KROM, N.J.	1926	'L'Art Javanais dans les Musées de Hollande et de Java', AA, VIII.
KRUIJT, J.A.	1877	Atjeh en de Atjehers. Leiden.
LAKING, G.F.	1964	Oriental Arms and Armour (Catalogue of the Wallace Collection). London.
LAIDLAW, G.M.	1947	'Some Notes on Keris Measurements', JMBBRAS (June Vol.).
LANGEN, K.F.H. VAN	1888	'Atjeh's Westkust', TKNAG.
LEENDERTZ, C.J.	1888	'Godsoordelen en Eeden', TKNAG, Tweede Serie, Deel V.
LEIGH, B.	1989	Tangan-Tangan Trampil, Seni Kerajinan Acheh / Hands of Time, The Crafts of Acheh. Jakarta.
LOEBÈR, J.A.	1915	Leder- en Perkamentwerk, Schorsbereiding en Aardewerk in Nederlandsch-Indië. Amsterdam.
	1916	Houtsnijwerk en Metaalbewerking in Nederlandsch-Indië. Amsterdam.
	1928	'Over de Soempitan in Indonesië en in Vlaanderen'. IG (December Vol.).
LORM, A.J. DE	1939	'Kantteekening bij eenige Mesheften uit Nias', CI, (Mei Nr.)
	1941a	'Zwaardgrepen en Mesheften van Nias', CI, Nr. 3.
	1941b	Indië in België: voorwerpen uit Nederlandsch Oost-Indië in Belgische openbare verzamelingen. Resultaten van een onderzoek in de jaren 1938 en 1939. Leiden.
	1942	'Een Merkwaardigheid van Zwaardamuletten van Zuid-Nias', CI, Nr. 3.
LOW, H.	1848	Sarawak, its Inhabitants and Productions. London.
MAATEN, K. VAN DER	n.d.	De Indische Oorlogen.
MARSDEN, W.	1811	The History of Sumatra. [Repr.: Kuala Lumpur, 1966]
MARYATT, F.S.	1848	Borneo and the Indian Archipelago. London.
MATTHES, B.F.	1874	Boegineesch-Hollandsch woordenboek, met Hollandsch-Boegineesche woordenlijst, en verklaring van een tot opheldering bijgevoegden ethnographischen atlas. 's-Gravenhage.
	1885	Makassaarsch-Hollandsch woordenboek, met Hollandsch-Makassaarsche woordenlijst, en verklaring van een tot opheldering bijgevoegden ethnographischen atlas. 's-Gravenhage.
MEIJER, J.J.	1916-1917	'Een Javaansch Handschrift over Pamormotieven', TNION.
MEYER, A.B. AND UHLE, M.	1885	Seltene Waffen aus Afrika, Asien und Amerika. Leipzig.
MILLER, E.Y.	1905	The Bataks of Palawan. Manila.
MJÖBERG, E.	1927	Borneo, het Land der Koppensnellers. Zeist.
MODIGLIANI, E.	1890	Un Viaggio a Nias. Milan.
MULLER, K.	1990	Oog op Indonesië. Hong Kong.
MÜLLER, F.W.K.	1893	'Beschreibung einer von G. Meissner zusammengestellten Batak-Sammlung', KMVV, III, 1/2.
MÜNSTERBERGER, W.	1939	'Die Ornamente an Dayak-Tanzschilden und ihre Beziehung zu Religion und Mythologie', CI, Band I.
MUSEUM OF ART, BOSTON	1949	Indonesian Art, a Loan Exhibition from the Royal Indies Institute Amsterdam, The Netherlands. Baltimore.
MUSEUM VOOR LAND- EN VOLKENKUNDE, ROTTERDAM	1965	Indonesië-Oceanië, kunst uit particulier bezit. Rotterdam.
MUSEUM VOOR VOLKENKUNDE, ROTTERDAM	1999	Decorative Arts of Sumba. Amsterdam/Kuala Lumpur.
NEWBOLD, T.J.	1839	Political and statistical account of the British settlements in the Straits of Malacca, viz. Pinang, Malacca, and Singapore; with a history of the Malayan states on the peninsula of Malacca. Vols. 1 & 2.
NIEUWENHUIZEN, W.C.	1897	'De Politiek van den Oorlog in Atjeh', IMT, Deel 2, Nr. 28.
OUDEMANS, A.C.	1889	'Engano, zijne Geschiedenis, Bewoners en Voortbrengselen', TKNAG, Tweede Serie, Deel VI.
PARAVICINI, E.M.M.	1923-1924	'Over Kapmessen van Nederlandsch Indië', TNION, 8e Jaargang.
PAREKH, K.	1971	'Deadly Beauty of the Kris', Orientations, 2.
PETRUS, J.TH.	1905-1906	'De Madoerees en zijn wapens', Weekblad voor Indië, 2, Afl. 4.
PLATENKAMP, J.D.M.	1990	De waarde der dingen. Ceremoniële geschenken der Tobelo. Den Haag.
PLEYTE, W.C.M.	n.d.	Sumpitan und Bogen in Indonesien.
	1892-1893	'Systematische Beschrijving van de door de Heeren Planten en Wertheim verzamelde Ethnographica tijdens hun Verblijf op de Zuidwester- en Zuidooster-eilanden', TNAG, IX, X.
	1894	'Systematische Beschrijving eener Ethnographische Verzameling, bijeengebracht ter Noordkust van Ceram', TNAG, Tweede Serie, deel XI.
PLUVIER, J.M.	1995	Historical Atlas of South-East Asia. Leiden.
RAFFLES, T.S.	1817a	The History of Java. Vol. I. London.
	1817b	The History of Java. Vol. II. London
RASSERS. W.H.	n.d.	On the Javanese Kris. BTLVNI, Deel 99, Afl. 4.
	1938	'Inleiding tot een Bestudeering van de Javaansche Kris', MKNAW, I, Nieuwe Reeks, Afd. Letterkunde.
RIJNDERS, H.W.M.J.	1985	Stille Krachten van de Kris. Den Haag.
ROBINSON, B.W.	1965	Exhibition Catalogue on Oriental Arts (The Tower of London). London
RODGERS, S.	1995	Power and Gold. Genève.
ROGERS, T.D.	1995	'A List of Names of North Sumatran Weapons and Hilts', SMJ, Vol. XLVIII, Nr. 69 (New Series).
ROSENHEIN, W.	1901	'Notes on Malay Metal-work', JRAIGBI, Nr. 31.
ROTH, W.L.	1896a	The Natives of Sarawak and British North Borneo. Vol. I. New York.
	1896b	The Natives of Sarawak and British North Borneo. Vol. II. London.
ROUFFAER, G.P.	1932	'Beeldende Kunst in Nederlandsch Indië', BKI, Nr. 89.
SCHEFOLD, R.	1979	Speelgoed voor de Zielen, Kunst en Cultuur van de Mentawei-eilanden. Delft.
SCHMELTZ, J.D.E.	1890	'Indonesische Prunkwaffen. Ein Beitrag zur Kunde des Kunstgewerbes in Indonesien und der ethnologischen Bedeutung der Kris', IAE, Band III.
	1893	'Über ein dajakisches und zwei japanische Schwerter', IAE, Band VI.
	1909	Cat. REM, Deel III, Catalogus der Bibliotheek, Leiden.
SCHRÖDER, C.A.	1859/1874	Ethnographische atlas bevattende afbeeldingen van voorwerpen uit het leven en de huishouding der Boeginezen; hoofdzakelijk dienende tot opheldering van het Boegineesch woordenboek van Dr. B.F. Matthes.
SCIDMORE, E.R.	1897	Java, The Garden of the East. New York.
SEITZ, H.	1938	Two Royal Kerisses. Ethnos, 3.
SELLATO, B.	1989	Hornbill and Dragon. Jakarta. [Repr. Singapore 1992].
SHELFORD, R.	1901	'A provisional Classification of the Swords of the Sarawak Tribes', JAI, Vol. XXXI.
SIBETH, A.	1990	Batak. Mit den Ahnen leben. Menschen in Indonesien. Stuttgart.
SIRAG, K.H.	1986	Katalogus van Indonesische wapens: collectie Bisseling. Driebergen.
	2000	Beschrijving van enkele zwaardtypen uit Indonesië. (Unpublished ms.). 's-Gravenhage.
SKEAT, W.W. AND BLAGDEN, C.O.	1906	Pagan Races of the Malay Peninsula. Vol. I. London.
SMITS, J.C.J.	1881	Gedenkboek van het Koloniaal-Militair Invalidenhuis Bronbeek. Arnhem.
SNELLEMAN, J.F.	1920-1921	'Houtsnijwerk van Groot-Kei', TNION.
SNOUCK HURGRONJE, C.	1892	'Notulen van de algemeene en bestuursvergaderingen van het Bataviaasch Genootschap van Kunsten en Wetenschappen', Afl. I, Deel XXX.
	1904	'Iets over Koedjang en Badi', TITLV, Deel XLVII.

SOEPARDI, P.	1939	Volkssmederijen en Cultuur in Java Instituut. Exhibition Catalogue Yogyakarta.	
SOLYOM, G. AND SOLYOM, B.	1978	The World of the Javanese Keris. Honolulu.	
SOPHER, D.E.	1965	'The Sea Nomads', MNMS, Nr. 5.	
SPAT, C.	1921	'Wapens der inlandsche bevolking', ENI, Tweede druk, IV.	
STAAT, D.	1996	'Insel-Schönheiten', Visier, Nr. 8 (August).	
STEINMANN, A.	1954	'Een Oud-Javaanse Kris met Voorstellingen uit de Mintagara', Djawa 2/3.	
	1966	'Figürliche Darstellungen als Verzierung javanischer Waffen', BSGAE, 42. Jahrgang.	
STINGL, H.	1969	'Schwerter aus Zentral-Kalimantan', JMVL, Band XXVI.	
STONE, G.C.	1934	A Glossary of the Construction, Decoration and Use of Arms and Armor. Portland.	
STRAATEN, H.S. VAN DER	1968a	'Krisvormen en Pamormotieven', VNN, 2e Jaargang, Nr. 2.	
	1968b	'Het Mysterie van de Kris', VNN, 3e Jaargang, Nr. 1.	
	1973	'De Occulte Wereld van de Kris', BRES, Nr. 30.	
STUTTERHEIM, W.F.	n.d.	Pictorial History of Civilisation in Java. Weltevreden.	
SUMINTARSIH	1990	Senjata tradisional Daerah Istemewa Yogyakarta. Proyek Penelitian Pengkajian dan Pembinaan Nilai-Nilai Budaya.	
SUNARTI	1993	Senjata tradisional Daerah Khusus Ibukota Jakarta. Proyek Inventarisasi dan Pembinaan Nilai-Nilai Budaya.	
SWETTENHAM, F.	1907	British Malaya. London/New York, 1907.	

T t

TAMMENS, G.J.F.J.	1991-1994	De Kris: Magic Relic of Old Indonesia. 3 Vols. Eelderwolde.
TAVARELLI, A.	1995	Protection, Power and Display: Shields of Island Southeast Asia and Melanesia. Boston.
TEEUW, A.	1996	Indonesisch-Nederlands Woordenboek. Leiden.
TEILLERS, J.W.	1910	Ethnographica in het Museum van het Bataviasch Genootschap voor Kunsten en Wetenschappen te Batavia. Weltevreden & 's-Gravenhage.
TROMP, S.W.	1888	'Mededeelingen omtrent Mandau's', IAE, I.
TUUK, H.N. VAN DER	1861	Bataksch-Nederduitsch woordenboek. Amsterdam.

U u

UHLMANN, W.	1999	Blankwaffen aus Ost- und Südost-Asien. Würzburg.

V v

VANVUGT, E.	1995	De schatten van Lombok. Amsterdam.
VELTMAN, T.J.	1904	'Nota betreffende de Atjehsche Goud- en Zilversmeedkunst', TITLV, XLVII.
VERMEIREN, P.	1981	'De Tjoendrik', Musket Info, Nr. 11.
	1982	'De Amanremoe', Musket Info, Nr. 18; VEMA, Jaargang 13, Nr. 2.
	1983a	'De Gliwang', Musket Info; VEMA, Jaargang 13, Nr. 4.
	1983b	'De Sikin', Musket Info, Nr. 23.
	1987	'Sabels van Noord Sumatra', Musket Info, Nr. 34.
VERSCHUER, F.H. VAN	1883	'De Badjo's', TKNAG, Nr. 7.
VIANELLO, G.	1966	Armi et Armature Orientali. Milano.
VOLZ, W.	1909	Nord-Sumatra. Band I. Berlin.
	1912	Nord-Sumatra. Band II. Berlin.
VOSKUIL, H.J.	1921-1922	'Over inheemsche Wapens in den Oost-Indischen Archipel', TNION, Zesde Jaargang.
	1931-1932	'Bij de foto's van Indonesische wapens uit de verzameling Wurfbain', TNION, 15.

W w

WAGNER, F.A.	1949	Sierkunst in Indonesië. Groningen/Batavia.
	1959	Indonesien. Die Kunst eines Inselreiches. Baden-Baden.
WALCOTT, A.S.	1914	Java and her Neighbours. New York.
WALUYO, H.	1993	Senjata Tradisional Daerah Bali. Proyek Penelitian Pengkajian dan Pembinaan Nilai Nilai Budaya.
WALLACE, A.R.	1890	The Malay Archipelago. London/New York.
WASSING-VISSER, R.	1995	Royal gifts from Indonesia. Historical bonds with the house Orange - Nassau (1600-1938). The Hague / Zwolle
WEIGLEIN, W. AND ZAHORKA, H.	1986	Expeditionen durch Indonesien. Frankfurt am Main.
WIGGERS, F.R. AND CARPENTER, B.W. (EDS.)	1999	Mentawai Art. Singapore.
WILKEN, G.A.	1893	Handleiding voor de vergelijkende volkenkunde van Nederlandsch-Indië naar diens dictaat en aanteekeningen uitgegeven door C.M. Pleyte. Leiden.
WINSTEDT, R.O.	1909	'Papers on Malay Subjects', Malay Industries, Part I.
	1912	'Three early Keris', JSBRAS, Nr. 62/63.
WIRAMAJA, L.	1991	Senjata Tradisional di Nusa Tenggara Barat. Departemen Pendidikan dan Kebudayaan.
WOODFORD, C.M.	1890	A Naturalist among the Head-Hunters. London.
WOOLLEY, G.C.	1938a	'Origin of the Malay Keris', JMBBRAS (Dec.).
	1938b	'Keris Measurements', JMBBRAS (Dec).
	1938c	'A New Book on the Keris', JMBBRAS (Dec).
	1947a	'Notes on Two Knives in the Pitt Rivers Museum', JMBBRAS (Dec).
	1947b	'The Malay Kris: Its Origin and Development', JMBBRAS, Vol. 20.

Z z

ZAAL, C. VAN DER	1936	Het leven van onze Toradja's. Dordrecht.
ZONDERVAN, H.	1888	'Timor en de Timoreezen', TKNAG, Tweede Serie, Deel V.
ZONNEVELD, A.G. VAN	1985	'De Indonesische Sikin Panjang', De Wapenverzamelaar, 22e Jaargang.
	1988	'De Indonesische Pandat', De Wapenverzamelaar, 24e Jaargang.
	1990	'De Javaanse Golok', De Wapenverzamelaar, 25e Jaargang.
	1994	'Een Classificatie van Indische Blanke Wapens I', De Wapenverzamelaar, 29e Jaargang.
	1995	'Een Classificatie van Indische Blanke Wapens II', De Wapenverzamelaar, 30e Jaargang.
	1996	De Wapens van Indië. Leiden.
	1997	'Een Classificatie van Indische Blanke Wapens III', De Wapenverzamelaar, 32e Jaargang.
	1998•	'Indische Wapens. Kunst, antiek of mystiek?', Origine, Jaargang 6, Nr. 1.

PROVENANCE OF THE OBJECTS AND PICTURES ILLUSTRATED

COLLECTIONS

MARION AND ERIC CRINCE LE ROY:
(PHOTOGRAPHY: F. HERREBRUGH)
Illus. C, 225, 226, 227, 228, 229, 230, 231, 232, 233, 234, 235, 236, 237, 238, 239, 240, 241, 242, 243, 244, 245, 246, 247, 248, 249, 250, 251, 252, 253, 254, 255, 256.

J. VAN DAALEN:
Illus. D, 216, 331, 552.

WILLEM VAN DER POST:
Illus. 7, 11, 19, 20, 31, 33, 34, 36, 37, 46, 72, 74, 83, 114, 132, 137, 140, 142, 146, 147, 149, 160, 164, 165, 167,174, 175, 180, 181, 183, 200, 201, 202, 204, 205, 206, 207, 209, 210, 212, 213, 217, 220, 222, 223, 258, 259, 263, 267, 277, 278, 279, 280, 284, 288, 289, 294, 319, 324, 333, 334, 335, 336, 337, 338, 365, 366, 369, 377, 384, 386, 387, 388, 390, 394, 406, 414, 415, 416, ,417, 418, 419, 420, 422, 427, 428, 429, 431, 432, 433, 434, 442, 445, 451, 456, 459, 460, 469, 470, 502, 507, 517, 525, 536, 543, 547, 550, 551, 559, 590, 601, 602, 606, 621, 629, 630, 631.

RIJKSMUSEUM VOOR VOLKENKUNDE/NATIONAL MUSEUM OF ETHNOLOGY (RMV):
(PHOTOGRAPHY: B. GRISHAAVER)
Illus. A, B, E, F, G, H, I, J, K, 1, 6, 18, 28, 38, 41, 48, 49, 50, 66, 68, 73, 75, 80, 81, 92, 95, 97, 103, 112, 115, 123, 126, 141, 161, 169, 179, 184, 186, 189, 208, 211, 214, 218, 219, 221, 224, 257, 260, 270, 272, 275, 276, 287, 290, 297, 298, 322, 332, 367, 368, 371, 391, 400, 421, 425, 436, 437, 457, 463, 492, 499, 512, 566, 592, 595, 597, 600, 603, 620, 625, 632, 633, 635, 636, 637, 641, 642.

KAREL SIRAG:
Illus. 5, 65, 67, 70, 84, 85, 94, 129, 153, 170, 171, 172, 173, 215, 265, 269, 273, 283, 286, 318, 323, , 352, 370, 379, 380, 383, 385, 392, 401, 405, 407, 408, 411, 447, 449, 500, 509, 526, 527, 528, 544, 554, 555, 556, 568, 569, 570, 571, 572, 573, 574, 575, 576, 577, 578, 579, 580, 581, 582, 583, 584, 585, 586, 587, 588, 589, 591, 628.

ALBERT VAN ZONNEVELD:
Illus. 2, 8, 9, 10, 15, 16, 21, 22, 23, 24, 25, 26, 32, 35, 60, 78, 87, 88, 99, 100, 101, 102, 104, 105, 116, 117, 118, 119, 120, 121, 122, 127, 130, 134, 135, 139, 143, 145, 148, 150, 151, 152, 154, 155, 156, 157, 158, 159, 166, 176, 185, 203, 261, 262, 264, 266, 268, 285, 299, 300, 301, 302, 308, 309, 314, 315, 316, 317, 326, 339, 340, 341, 342, 343, 344, 378, 389, 393, 395, 396, 397, 402, 409, 410, 412, 413, 444, 446, 448, 450, 455, 458, 461, 462, 465, 471, 472, 473, 474, 475, 476, 477, 478, 479, 503, 504, 505, 506, 508, 510, 511, 513, 514, 515, 516, 518, 519, 520, 523, 538, 548, 549, 557, 558, 560, 561, 562, 563, 564, 565, 599, 613, 615, 622, 646, 647.

PUBLICATIONS

BEZEMER, T.J. - Indonesische Kunstnijverheid. 's-Gravenhage, 1931.
Illus. 522, 624.

BISSELING, G.A.L. - Tekeningen van Indonesische Wapens. Unpublished ms. (c.1950).
Illus. 3, 4, 62, 63, 64, 93, 96, 98, 113, 281, 282, 292, 350, 398, 404, 423, 424, 498, 545, 627.

CLERCQ, F.S.A. DE - Bijdragen tot de kennis der residentie Ternate. Leiden, 1890.
Illus. 489.

FISCHER, H.W. - De Eilanden om Sumatra. (Cat. REM IV). Leiden, 1909.
Illus. 296, 372, 373.

FISCHER, H.W. - Bataklanden, Sumatra II. (Cat. REM, VIII). Leiden, 1914.
Illus. 443, 466, 467.

FOY, W. - 'Schwerter von der Celébes See' (PKEMD, Band XII). Dresden, 1899.
Illus. 111.

GARDNER, G.B. - Keris and Other Malay Weapons. Singapore, 1936. (Repr.: Wakefield, 1973).
Illus. 39, 58, 311, 382, 454, 611.

GRUBAUER, A. - Unter Kopfjägern in Central-Celebes. Leipzig, 1913.
Illus. 125, 345, 487, 488, 490, 491, 493, 494, 495, 496, 604, 605, 639.

HEIN, A.R. - Die Bildende Künste bei den Dayaks auf Borneo. Wien, 1890.
Illus. 89, 596, 351, 353, 483.

HEYST, A.F.C.A.VAN - 'Anggang Gading' (CI, Vol. 4, Jan./Feb). 1942.
Illus. 12, 13.

HILL, A.H. - 'The Keris and other Malay Weapons' (JMBRAS, Vol., 29, Part 4, No. 176). 1956.
Illus. 194, 195.

JACOBSEN, A. - Reise in die Inselwelt des Banda-Meeres. Berlin, 1896.
Illus. 69, 271, 349, 521, 524, 640.

JASPER, J.E. and PIRNGADIE, M. - De Inlandsche Kunstnijverheid in Nederlandsch Indië. Deel 5. Den Haag, 1930.
Illus. 45, 76, 77, 196, 197, 198, 199, 312, 313, 330, 347, 452, 453, 468, 610, 644.

JONGE, N. DE and DIJK, T. VAN - Tanimbar. De unieke Molukken-foto's van P. Drabbe. Amsterdam, 1995.
Illus. 567.

KOL, H.H. VAN - Driemaal dwars door Sumatra en zwerftochten door Borneo. Rotterdam, 1914.
Illus. 42, 497.

KREEMER, J. - Atjèh. I. Leiden, 1922.
Illus. 348, 626.

MARSDEN, W. - The History of Sumatra. 1811. [Repr.: Kuala Lumpur, 1966].
Illus. 108, 168.

MJÖBERG, E. - Borneo, het Land der Koppensnellers. Zeist, 1927.
Illus. 546.

MODIGLIANI, E. - Un Viaggio a Nias. Milan, 1890.
Illus. 30, 40, 61, 79, 90, 91, 354, 355, 356, 357, 358, 359, 360, 361, 362, 363, 364, 594, 608, 609, 612, 614, 638.

MÜLLER, F.W.K. - 'Beschreibung einer von G. Meissner zusammengestellten Batak-Sammlung' (KMVV, III, 1/2) 1893.
Illus. 124, 307, 320, 403.

PARAVICINI, E.M.M. - 'Over Kapmessen van Nederlandsch Indië' (TNION, 8e Jaargang). 1923-1924.
Illus. 71, 291, 295, 643.

RAFFLES, T.S. - The History of Java. Vol. I. London, 1817.
Illus. 14, 43, 59, 82, 86, 106, 107, 162, 293, 329, 346, 399, 598, 607, 645.

ROTH, W.L. - The Natives of Sarawak and British North Borneo. Vol. I. New York, 1896.
Illus. 47, 464, 480.

ROTH, W.L. - The Natives of Sarawak and British North Borneo. Vol. II. London, 1896.
Illus. 27, 109, 110, 163, 182, 187, 188, 190, 191, 310, 375, 376, 484, 485, 486, 617, 618, 619, 623, 634.

TUUK, H.N. VAN DER - Bataksch-Nederduitsch woordenboek. Amsterdam, 1861.
Illus. 321, 426, 430, 435, 553.

VOLZ, W. - Nord-Sumatra. Band I. Berlin, 1909.
Illus. 303, 304, 305, 306, 381, 438, 439, 440, 441, 481, 529, 530, 531, 532, 533, 534, 535, 537, 539, 540, 542, 593.

VOLZ, W. - Nord-Sumatra. Band II. Berlin, 1912.
Illus. 128, 131, 133, 136, 138, 144, 274, 325, 327, 328, 482, 501.

TRADITIONAL WEAPONS OF THE INDONESIAN ARCHIPELAGO   **TRADITIONAL WEAPONS** OF THE INDONESIAN ARCHIPELAGO   **TRADITIONAL WEAPONS** OF THE INDONESIAN ARCHIPELAGO